Instrumentation

TUTORIAL GUIDES IN ELECTRONIC ENGINEERING

Series editors
Professor G.G. Bloodworth, *University of York*
Professor A.P. Dorey, *University of Lancaster*
Professor J.K. Fidler, *University of York*

This series is aimed at first- and second-year undergraduate courses. Each text is complete in itself, although linked with others in the series. Where possible, the trend towards a 'systems' approach is acknowledged, but classical fundamental areas of study have not been excluded. Worked examples feature prominently and indicate, where appropriate, a number of approaches to the same problem.

A format providing marginal notes has been adopted to allow the authors to include ideas and material to support the main text. These notes include references to standard mainstream texts and commentary on the applicability of solution methods, aimed particularly at covering points normally found difficult. Graded problems are provided at the end of each chapter, with answers at the end of the book.

1 Transistor Circuit Techniques: discrete and integrated (2nd edition) — G.J. Ritchie
2 Feedback Circuits and Op Amps (2nd edition) — D.H. Horrocks
3 Pascal for Electronic Engineers (2nd edition) — J. Attikiouzel
4 Computers and Microprocessors: components and systems (2nd edition) — A.C. Downton
5 Telecommunication Principles (2nd edition) — J.J. O'Reilly
6 Digital Logic Techniques: principles and practice (2nd edition) — T.J. Stonham
7 Instrumentation: Transducers and Interfacing (new edition) — B.R. Bannister and D.G. Whitehead
8 Signals and Systems: models and behaviour — M.L. Meade and C.R. Dillon
9 Basic Electromagnetism and its Applications — A.J. Compton
10 Electromagnetism for Electronic Engineers — R.G. Carter
11 Power Electronics — D.A. Bradley
12 Semiconductor Devices: how they work — J.J. Sparkes
13 Electronic Components and Technology: engineering applications — S.J. Sangwine
14 Optoelectronics — J. Watson
15 Control Engineering — C. Bissell
16 Basic Mathematics for Electronic Engineers: models and applications — Szymanski
17 Software Engineering — D. Ince
18 Integrated Circuit Design and Technology — M.J. Morant

Instrumentation: Transducers and Interfacing

New edition

B.R. Bannister and D.G. Whitehead
Department of Electronic Engineering
University of Hull

CHAPMAN AND HALL

University and Professional

LONDON·NEW YORK·TOKYO·MELBOURNE·MADRAS

UK	Chapman and Hall, 2–6 Boundary Row, London SE1 8HN
USA	Van Nostrand Reinhold, 115 5th Avenue, New York NY10003
JAPAN	Chapman and Hall Japan, Thomson Publishing Japan, Hirakawacho Nemoto Building, 7F, 1-7-11 Hirakawa-cho, Chiyoda-ku, Tokyo 102
AUSTRALIA	Chapman and Hall Australia, Thomas Nelson Australia, 480 La Trobe Street, PO Box 4725, Melbourne 3000
INDIA	Chapman and Hall India, R. Seshadri, 32 Second Main Road, CIT East, Madras 600 035

First edition 1986, first published as *Transducers and Interfacing — principles and techniques*
New edition 1991

© 1991 B.R. Bannister and D.G. Whitehead

Typeset in 10/12pt Times by KEYTEC, Bridport, Dorset
Printed in England by Clays Ltd, St. Ives PLC

ISBN 0 412 34240 5   0 442 30514 1 (USA)

British Library Cataloguing in Publication Data
Bannister, B. R. (Brian Roy) 1936–
   Instrumentation: transducers and interfacing.
   1. Electronic equipment. Transducers
   I. Title II. Whitehead, D. G. (Donald Gill) 1937– III.
   Series
   621.381532

   ISBN 0-412-34240-5

Library of Congress Cataloging-in-Publication Data

available

Contents

Preface to first edition

It is true that transducer operation and interfacing can be defined as peripheral activities in the sense that they form the links between the purely electronic system, or circuit, and the external world. It is unfortunately true that many engineers also tend to consider these areas as peripheral to the main body of systems design, whereas, in fact, they play an increasingly central part in any engineering activity.

It is our intention in this book to introduce the reader to the basic techniques involved when electronic systems are to interact with the 'real world'. Interaction here covers the entire range from the collection of data using sensing transducers, through the transmission of the data and its conversion to other more convenient forms, and, finally, to the control of output transducers (actuators) and the display of information.

The successful application of electronics to measurement and control necessitates an appreciation both of transducer operation and of methods of ensuring their correct functioning in particular circumstances. This book covers both aspects — transducers and interfacing — and, though intended primarily for students at first or second year level of degree or diploma courses in electrical and electronic engineering, it will also be found useful by students in many related fields of engineering and science. Wherever possible, practical details and examples based on modern devices are included.

The first two chapters are concerned with the tranducers themselves; basic principles, performance criteria and limitations are introduced in Chapter 1, and more practical applications are considered in Chapter 2. The next two chapters deal with the processing of the signals produced by the transducers; Chapter 3 discusses the operational amplifier in detail and describes the more specialized types of circuit used in instrumentation amplifiers. Amplitude and frequency modulation techniques and the analogue scanner are also included. Although most transducers are analogue in nature, the means of processing information nowadays is almost always digital, and Chapter 4 is therefore devoted to analogue-to-digital and digital-to-analogue conversion methods. The general principles of the most popular types of converter are explained, and important parameters defined. Sample-and-hold devices and voltage-to-frequency converters are also considered in this chapter. The fifth and final chapter is concerned with interfacing devices and sub-systems, both analogue and digital, and methods of dealing with the problems that are likely to be encountered, especially those involving electrical noise. The interfacing methods used in microprocessor-based systems are reviewed, and common international standards for data logger and telemetry systems are introduced.

In our attempt to combine a thorough treatment with a broad perspective in a subject area as loosely defined as this, it has been invaluable to have had the assistance and advice of the series editors throughout the preparation of the book. Our particular thanks go to Professor Kel Fidler, of the Open University, for his tactful suggestions and corrections and ensuring that we did not stray too far from our original intentions.

Preface to new edition

It is true that transducer operation and interfacing can be defined as peripheral activities in the sense that they form the links between the purely electronic system, or circuit, and the external world. It is unfortunately true that many engineers also tend to consider these areas as peripheral to the main body of systems design, whereas in fact they form an increasingly important part of any engineering activity.

This book provides an introduction to the basic techniques involved in modern instrumentation, in which electronic systems interact with the 'real world'. Interaction here covers the entire range of activities from the collection of data using sensing transducers, through the transmission of the data and its conversion to other more convenient forms, and finally, to the control of output transducers (actuators) and the display of information.

The successful application of electronics to instrumentation systems necessitates an appreciation both of transducer operation and of methods of ensuring their correct functioning in particular circumstances. This book covers both aspects – transducers and interfacing – and, though intended primarily for students at first or second year level of degree or diploma courses in electrical and electronic engineering, it will also be found useful by students in many related fields of engineering and science. In preparing this new and enlarged edition we have retained the approach adopted in the original edition, including wherever possible practical details and examples based on modern devices.

The first two chapters are concerned with the transducers themselves; basic principles, performance criteria and limitation are introduced in Chapter 1, and more practical applications are considered in Chapter 2. The next 3 chapters deal with the processing of the signals produced by the transducers; Chapter 3 discusses the operational amplifier in detail and describes the more specialized types of circuit used in instrumentation amplifiers. Amplitude and frequency modulation techniques and the analogue scanner are also included. Whilst the majority of transducers are analogue in nature, the means of processing information at present is almost always digital, and Chapter 4 is thus devoted to analogue-to-digital and digital-to-analogue conversion methods. The general principles of the most popular types of converter are explained, and important parameters defined. Sample-and-hold devices and voltage-to-frequency converters are also considered in this chapter. The fifth chapter is entirely new and introduces the principles which form the basis of the processing of signals in the digital domain. The sixth and final chapter is concerned with interfacing devices and sub-systems, both analogue and digital, and methods of dealing with the problems which are likely to be encountered. The interfacing methods used in microprocessor-based systems are reviewed and common international standards for data logger and telemetry systems are introduced.

In our attempt to combine a thorough treatment with a broad perspective in a subject area as large as this, it has been invaluable to have had the assistance and advice of the series editors throughout the preparation of this new edition. Out particular thanks go to Professor Kel Fidler and Professor Greville Bloodworth, both of York University, for their tactful suggestions and corrections and ensuring that we did not stray too far from our original intentions.

1
Principles of transduction

Objectives

☐ To introduce the basic concepts and terminology
☐ To consider the practical limitations of transducers, in terms of
    accuracy   reliability
    hysteresis   repeatability
    linearity   sensitivity
    predictability.
☐ To indicate failure mechanisms and to calculate mean failure times.
☐ To review the essential physical theory relating to the operation of common transducers.

In the monitoring and control of processes and operations, a multitude of devices are used to collect or present information in a suitable form; such devices are termed **transducers**. The Oxford Dictionary defines a transducer as a device which accepts energy from one part of a system and emits it in different form to another part of the system. We are concerned with electronic instrumentation systems and the devices that we will consider are those which either, by monitoring the physical world, provide an electrical analogue to be used in the processing or display system, or, by responding to an electrical stimulus in an ordered manner, interact with the physical world. In some cases the required transduction may involve more than one process. An accelerometer, for instance, may use a strain gauge (the primary sensor) to measure the strain in a steel bar as it flexes under the forces of acceleration. Then, since strain is proportional to stress, that is to force per unit area, this indicated value gives us the quantity we really wish to measure.

Young's modulus, $E = $ stress/strain; see Chapter 2.

A transducer whose output energy is supplied entirely by its input signal is termed passive. A ceramic cartridge used in a hi-fi record player pickup is a passive transducer since it utilizes a piezoelectric effect to generate voltages proportional to the mechanical stress imparted by the stylus movement in the record groove. An active transducer has an external source of energy which supplies most of the output power. A temperature detector using a thermistor, for example, as we shall see, has a constant external voltage to drive current through the thermistor as it varies in resistance in response to temperature changes.

See p. 12.

The system of Fig. 1.1 illustrates the sort of arrangement we commonly encounter. Two distinct types of transducer are shown; the first we shall call the input transducer, or **sensor**, because it accepts signals from the process being monitored. Since in the 'real world' the vast majority of variables are continuous functions, these signals are usually in analogue form, meaning that they can assume any value between upper and lower limits. The output from

Any measuring device which extracts power from the system it is monitoring inevitably introduces errors since the measured quantity is disturbed by the act of measurement. High-quality transducers are designed to minimize this disturbance but it is always there in some degree.

Fig. 1.1 Process control

the transducer is then conditioned, because, although it is an electrical signal, it may not be in a form suitable for the monitoring system. It may, for example, have to be amplified or be converted to a digital form, in which the coded signal can exist only at certain defined levels. The second type is the output transducer which is designed to accept an electrical signal from the monitoring system and convert it into a form suitable for the 'real world'. The transducer could simply be an indicating lamp, converting the electrical signal into a visual one; this is a data presentation device. Alternatively, it might be necessary for the monitoring system to generate a mechanical output, giving rotational or linear motion. Such transducers would include valves, solenoids, pumps and motors, and are generally referred to collectively as **actuators**.

There are several important considerations to be taken into account in choosing a sensor for a particular application. These include:

sensitivity   repeatability
accuracy   reliability
linearity   cost

and, in addition, it is necessary to know the limits between which the information provided by the sensor is valid.

In general terms we can consider a sensor to have several input components as shown in Fig. 1.2. These components are

the desired signal, S_d
a noise signal, S_n
a modifying signal, S_m

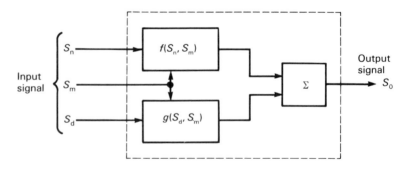

Fig. 1.2 Signal components

The functions $f(S_n, S_m)$ and $g(S_d, S_m)$ describe the conversion process, as yet undefined, that transforms the input signal into the appropriate form for the output. Quite simply, the functions, f, g, indicate that the output signal is some function of the input stimulus, and is proportional to the stimulus in some way. The noise input represents any unwanted signals to which the sensor is responsive. A magnetic cartridge of a record player, for example, is sensitive not only to the vibration of the stylus, as intended, but also to the alternating magnetic field of the nearby mains transformer, giving rise to mains hum. The modifying stimulus, S_m, represents external signals that can cause the response of the transducer to alter. These signals affect both the true signal, S_d and the unwanted signal, S_n. The moving-coil ammeter provides an example; a variation in the torsion constant of the springs controlling the position of the pointer would modify the readings of the meter. A noise stimulus in this context could be introduced by the physical position of the meter. If not placed horizontally, many instruments with pointers not precisely counter-balanced will indicate a bias, giving an apparent reading when no true signal is present. This bias is also modified by the strength of the springs.

$I = K/BAN$ where K is the torsion constant, B is the flux density, A is the area cut by the flux and N is the number of turns.

The **sensitivity** of a transducer is defined as the rate of change of the output signal, S_o, as the desired signal at the input varies. In mathematical form,

$$\frac{dS_o}{dS_d} = K + \frac{dS_n}{dS_d} + \frac{dS_m}{dS_d}$$

When a range of measurements is made of any process, it is essential to know the accuracy of the readings, and whether the accuracy is maintained over the entire range. As we noted earlier, any measuring device inevitably disturbs the process being monitored, and so we can never, in fact, obtain 'true' readings. However we can manage with the actual readings if we know to what accuracy they were made, and, for that reason, measurements should always be quoted with reference to the accuracy obtainable. It follows then that the characteristics of the measuring system should be known precisely, and here we must be very careful with our definitions. We cannot define in terms of absolute error, because the error, being the difference between the actual reading and the 'true' value, cannot be calculated if the true value is not known! Also in many cases it is not the accuracy of the transducer alone which is important. The relationship between stimulus and output is crucial to the measurement process, and we must consider the system as a whole. In taking even a single measurement we are subject to a statistical process, involving errors introduced by every component of the system, including the observer, and all we can do is to impose bounds on the overall error. Accuracy, therefore, relates to the precision of the measurement process. The ability to maintain accuracy over the entire measurement range is governed by the linearity of the system, that is, how nearly the response is directly proportional to the stimulus. The graph of Fig. 1.3 presents a set of readings taken with some increment over a range of values of the stimulus. Statistical scattering of the readings occurs, and whether we can describe the response as linear depends on how great a precision we demand. By accepting an error bound which ensures that all readings are within the parallel lines we can confidently state that the response is linear, but this is meaningful only if the

3

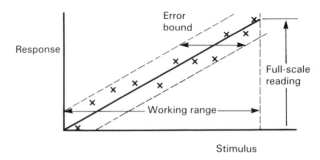

Fig. 1.3 System linearity

error bound is sufficiently small. A common method of specifying a linear transducer is by reference to an average calibration curve, which is taken as the straight line that best fits the scattered set of readings. The criterion used is generally that of least-squares fit. In this approach, the calibration line is drawn so that the sum of the squares of the vertical differences, between the readings and the line, is minimized.

In engineering practice, the **accuracy** of an instrument as defined in terms of the greatest horizontal deviation from the calibration line, is commonly quoted as a percentage of the full-scale reading. Thus if a voltmeter with a range of 0 to 10 V has a quoted accuracy of $\pm 1.0\%$ of fsd (full-scale deflection), quoted in this way it can be assumed that no error greater than ± 0.1 V will occur over the entire working range of the instrument.

Note that accuracies quoted in this way can be deceptive; a measurement of 1 V, say, could be in error by 10%!

In many applications the absolute linearity of the transducer is not important. Subsequent conditioning of the signal can make appropriate adjustments if necessary. In these circumstances it is the predictability and reliability of the performance which are important. We can cope with the non-linearity if we know exactly how the transducer will respond, and that it will do so reliably over many operations. In some cases a highly non-linear response is actually desirable. This is particularly true of the thermostat which is widely used and comes in many guises. The bi-metal snap-action switch, for example, is a simple device made from two metals, with different expansion coefficients, formed into a disc (Fig. 1.4(a)). At a certain temperature the

This point is often overlooked by designers who overspecify the linearity requirements and so increase the cost of a system.

(a)

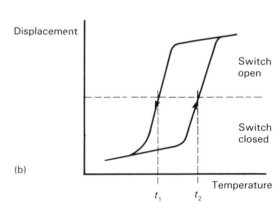

(b)

Fig. 1.4 The bi-metal disc switch

mechanical stress produced by the unequal expansion causes the disc to 'snap' into the concave shape, opening or closing electrical contacts similar to a microswitch. Transducers operating on this principle respond to the change in analogue signal, usually temperature, with a discrete change in physical shape. The graph of displacement against temperature for the bi-metallic disc is shown in Fig. 1.4(b) and is certainly not linear, having a pronounced hysteresis effect. In general terms, hysteresis represents the loss of energy associated with a physical process, and, in this case, is manifested in the two temperatures t_1 and t_2 at which the switching action takes place. For an accurate transducer, this hysteresis effect should be as small as possible.

Hysteresis can be undesirable; static friction, 'stiction', in a meter movement, for example, can cause inaccurate readings.

In addition to the accuracy of a transducer, an important factor is its **reliability**, since we would expect the transducer to generate virtually identical responses to the same input stimulus over its entire working life. In fact, we can turn this statement around and state that the working life of a transducer will be defined as that period of time over which it continues to perform accurately (within predetermined limits). Reliability is often related to the cost of the device, and the specification of working life then becomes an exercise in cost-effectiveness.

Failure of any component may be sudden and not capable of prediction, or gradual, in which case it might be possible to detect some movement out of specification. Furthermore, when failure occurs it could be complete or only partial. Failures which are both sudden and complete are said to be **catastrophic**, whereas those that are partial and gradual are **degradation** failures. Failure of any nature can occur because of inherent weakness in the device, or it can be induced by operating the transducer outside its designed capabilities.

The assessment of reliability is normally quoted by manufacturers in statistical terms since it is usually impracticable to measure all parameters of every transducer. The reliability is then, in effect, the probability of the transducer performing satisfactorily within specification: it is the **confidence factor**. Eventually, all devices fail and the lifetime failure pattern of a batch of identical devices can be summarized in a diagram known as the bath-tub diagram because of its shape (Fig. 1.5). It has three distinct regions: early failure, constant failure and wearout failure. The early failure period is often called the burn-in period and, where reliability is very important, can be imposed on the devices by operating them under appropriate conditions for a

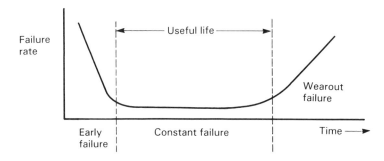

Fig. 1.5 The 'bathtub' diagram

sufficient length of time to remove weak elements, before the remainder are put to use in a system. The useful life extends over the constant failure period, and the aim is to reduce failures in this region to as low a level as possible. The time period during which devices operate satisfactorily varies widely from one to another, and all we can do is to give some indication of the average time over which the operation will be satisfactory.

The **mean time to failure** (MTTF) is defined over a total period of use as

$$\frac{\{\text{lifespan of each failure}\} + \{(\text{number of survivors}) \times (\text{period of use})\}}{\text{total number of failures}}$$

Worked Example 1.1

Over a 1000-hour period, 27 transducers of type X worked satisfactorily throughout, one failed after 680 hours, another failed after 13 hours and another failed after only 6 hours. What is the MTTF?

The mean time to failure is

Note that the MTTF is normally much greater than the period of use, or test.

$$\text{MTTF} = \frac{(1 \times 6) + (1 \times 13) + (1 \times 680) + (27 \times 1000)}{3} = 9233 \text{ hours.}$$

Where a device is repairable, the figure quoted is often the **mean time between failures** (MTBF), and is defined as

$$\text{MTBF} = \frac{\text{total period of use of the devices (device hours)}}{\text{number of failures}}$$

Worked Example 1.2

Assume transducer type X is repairable, so the three failures listed above were repaired, but one of them failed once more during the 1000 hours of use. What is the MTBF?

The mean time between failures is

$$\text{MTBF} = \frac{30 \times 1000}{4} = 7500 \text{ hours.}$$

Underlying physical principles of transducers

The operation of most transducers can be explained in terms of only a small number of basic principles, which is rather surprising when we consider the plethora of sensors, displays and actuators available and, for the remainder of this chapter, we will consider some of the underlying principles which are exploited in the design of modern transducers.

The electromechanical switch

An electromechanical sensor is one in which mechanical movement in some form gives rise to an electrical signal. The simplest form is the switch, where a mechanical movement opening or closing a contact will make or break an electrical circuit. This is a digital sensor since it is a two-state device, either 'off' or 'on'. More sophisticated sensors to indicate displacement, either rotary or translational, can be constructed using this elementary switching action. In the tachometer encoder several metal segments arranged around a disc (or linearly on a flat surface for translational sensing) are used with a metallic 'finger' contact to create a series of switches. Movement of the segments makes or breaks contact with the finger and the resulting series of electrical pulses, if counted, gives a measure of the distance moved. Provided the motion is in one direction only a single set of segments is adequate but to sense reversible movement two tracks are needed, so arranged that the waveform from the second is displaced by one quarter-cycle relative to the first (Fig 1.6). The phasing of the waveforms is governed by which contact makes or breaks first and is used to indicate the direction of travel. Incremental encoders of this sort often have a third, datum, track with a single segment providing a start or zero reference point.

Angular position sensing is usually achieved by including on the disc several sets of contacts - arranged as shown in Fig. 1.7 to give a direct binary indication. If we consider a 'closed' contact as signifying a logical 'one', and an 'open' contact a logical 'zero', then it can be seen that, as the disc is rotated, so all possible combinations of the binary codes, from 000 to 111, will be generated by the switches during one revolution of the disc. Unfortunately, if segment boundaries happen to fall under the contacts, it is very likely that mechanical misalignment of the contacts will lead to false readings of some of the bits and, with the standard binary code illustrated, the errors can be such as to make the encoder useless. In order to overcome these problems, the code used is invariably a Gray code in which the code for each segment differs from its neighbours in only one bit. Misreading of a single bit then leads to a maximum error of ±1 segment. The problems associated with the use of sliding mechanical contacts are very great, with dirt, vibration and the inevitable wear making for unreliable operation. Most high-resolution shaft encoders, therefore, use optical switches which differentiate between

Gray codes were introduced by Dr F Gray, (US Patent No. 2632058, March 1953). See *Fundamentals of Modern Digital Systems* or *Digital Logic Techniques.*

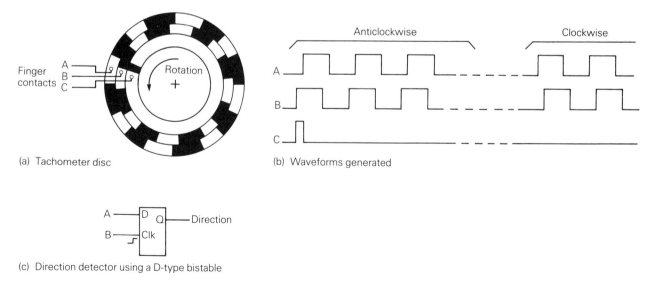

(a) Tachometer disc

(b) Waveforms generated

(c) Direction detector using a D-type bistable

Fig. 1.6

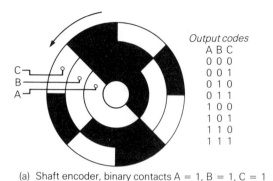

Output codes
A B C
0 0 0
0 0 1
0 1 0
0 1 1
1 0 0
1 0 1
1 1 0
1 1 1

(a) Shaft encoder, binary contacts A = 1, B = 1, C = 1

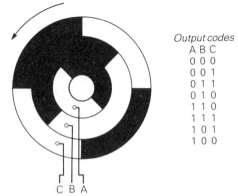

Output codes
A B C
0 0 0
0 0 1
0 1 1
0 1 0
1 1 0
1 1 1
1 0 1
1 0 0

(b) Shaft encoder, Gray contacts A = 1, B = 1, C = 1

(c) Gray to binary converter using ex-or gates

Fig. 1.7 The shaft encoder

transparent and opaque segments, rather than conducting segments. The discs are produced photographically, and, for those willing to pay, optical shaft encoders can be obtained with as many as twenty switches, giving a resolution of one part in 2^{20}. Lower resolution encoders, including tachometer encoders, often employ magnetic sensors which utilize the Hall effect. (The Hall effect is covered later in this chapter.)

Resistance

Resistance transducers rely on the flow of current to generate a voltage. A simple potentiometer, Fig. 1.8, is a tapped resistor allowing a lower voltage to be derived from a higher. The output voltage, V_o is directly related to the applied voltage, V_i, and $V_o = V_i R_2/(R_1 + R_2)$. By noting the output voltage for any given position of the tapping point, we can quantify linear or rotational movement. Rotary movement can be quantified using a simple

Ohm's law: $V = IR$.
The same current I flows in both parts of the resistor, so $V_0 = IR_2$. $V_i = I(R_1 + R_2)$.

$$V_o = V_i \frac{R_2}{R_1 + R_2}$$

Fig. 1.8 The simple potentiometer

8

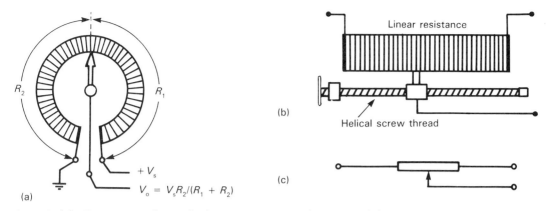

Fig. 1.9 (a) Representation of the rotary potentiometer. (b) The helical screw, multi-turn potentiometer. (c) Schematic representation

rotary potentiometer, Fig. 1.9. The resistance element is traditionally a closewound coil of resistance wire such as nichrome. The resolution of such potentiometers is defined as the average increment of output. This is governed by the number of turns of resistance wire making up the resistive element. Resolution is therefore expressed, as a percentage, as $100/$(number of turns). High-resolution potentiometers (better than $\pm0.01\%$) have a helical screwthread along which a wiper contact moves with, typically, ten turns of the screw being required to move the wiper contact from one end of its travel to the other. Linearity is also an important parameter in this context, and can be defined as the maximum deviation from the expected response, which in this case is a linear relationship between electrical output and mechanical travel. For a high-quality wirewound potentiometer a linearity of $\pm0.25\%$ is typical. The change in resistance as temperature varies is not so important, as these changes affect both sections of the element proportionally. However, if the parameter being sensed is transduced into a resistance change then clearly temperature dependence is a crucial factor. We define the temperature coefficient of resistance as

$$\frac{\delta R}{R}\frac{1}{\delta T}\ \Omega/\Omega\,^{\circ}\mathrm{C}^{-1}$$

where R is the resistance at the initial temperature, δR is the change in resistance over the temperature range δT. For wirewound potentiometers, a coefficient of around 20 ppm/°C is typical.

For greater resolution, Cermet resistance elements are available, using a metallized ceramic substrate combining high resolution with low temperature coefficient. More recently, conductive plastic film has become popular for high-quality potentiometers since, although it has a higher temperature coefficient than Cermet devices (typically 200 ppm/°C), the greater reliability, high resolution and low noise of the hard mirror-finish plastic film give definite advantages. An important feature of the plastic film resistor is that the resistive element can be trimmed, either by laser or by computer-controlled milling, to allow precise matching of an element to a linear or non-linear function.

ppm is parts per million.

9

A rotary transducer is constructed from a wirewound potentiometer of resistance 1000 Ω. If the output of the transducer is taken to a voltage display device which has an input resistance of 10 kΩ, derive an expression for the non-linearity of the circuit due to the loading effect of the display device. Hence find the maximum displayed error.

Considering only the potentiometer, if R_p is the potentiometer resistance and x is the fractional displacement, we have an output voltage

$$V_o = \frac{V_s R_p x}{R_p} = V_s x$$

Thus, as expected, the output voltage varies linearly with fractional displacement of the wiper. The Thevenin equivalent circuit gives

$$V_{th} = V_o = V_s x$$

and

$$R_{th} = (R_p x) \text{ in parallel with } [R_p(1-x)]$$
$$= \frac{(R_p x)[R_p(1-x)]}{R_p} = R_p x(1-x)$$

When the load is applied, the output becomes

$$V_o^* = \frac{V_{th} R_L}{R_{th} + R_L} = \frac{V_s x R_L}{R_p x(1-x) + R_L}$$

Now the correct output should be V_o, so the percentage error is

$$E = \frac{V_o^* - V_o}{V_o} \times 100\%$$

This is

$$E = \left(\frac{V_s x R_L}{R_p x(1-x) + R_L} - V_s x \right) \Big/ V_s x \times 100\%$$

$$= \left(\frac{R_L}{R_p x(1-x) + R_L} - 1 \right) \times 100\%$$

$$= \left(\frac{1}{\dfrac{R_p}{R_L} x(1-x) + 1} - 1 \right) \times 100\%$$

It can be seen that if R_L is much larger than R_p, the error approaches zero. The error also approximates to zero as x approaches 0 or 1. By differentiating and setting dE/dx to zero, it is seen that the maximum error occurs when $x = 0.5$. Thus the maximum error is

$$E_{max} = \left(\frac{10\,000}{(1000 \times 0.5 \times 0.5) + 10\,000} - 1 \right) \times 100\%$$

$$= -2.4\%$$

The simple Wheatstone bridge, Fig. 1.10, used in accurate measurement of

Fig. 1.10 The Wheatstone bridge

resistance, is, in effect, two potentiometers in parallel, with the tapping points at balance adjusted to give equal voltages. In general terms,

$$V_o = V_i \left[\frac{R_3}{R_3 + R_4} - \frac{R_1}{R_1 + R_2} \right]$$

and, for V_o to be zero,

$$\frac{R_3/R_4}{1 + R_3/R_4} = \frac{R_1/R_2}{1 + R_1/R_2}$$

whence

$$R_1/R_2 = R_3/R_4$$

The actual value of a resistive element is affected by many parameters. In terms of linear dimensions, the resistance of a conductor is given by

$$R = \rho L/A$$

where ρ is the resistivity, L is the length and A is the cross-sectional area.

For a very small change in length, where we can assume that the cross-sectional area does not change, dR is proportional to dL, though the change in resistance is also very small. In fact, any change in length leads also to a change in cross-sectional area and to a change in resistivity. Cross-sectional area changes are defined in terms of Poisson's ratio, v, which is the ratio of the transverse strain to the longitudinal strain causing it. For most materials this has a value between -0.25 and -0.4. The strain-induced resistivity change is known as the piezoresistive effect, which in metals is very small but in semiconductors is two orders of magnitude greater. Variations of resistance with temperature are governed by the relationship

$$R_T = R_o(1 + \alpha_1 T + \alpha_2 T^2 + \ldots)$$

where R_o is the resistance at $0\,°C$, R_T is the resistance at $T\,°C$ and α_1, α_2, etc. are the temperature coefficients of resistance.

In the case of metals, all coefficients other than α_1 are negligibly small and the resistance changes approximate to a linear function, $R_T = R_o(1 + \alpha T)$. As we shall see in Chapter 2, there are resistance devices, thermistors, with highly non-linear temperature characteristics and, normally, an overall negative temperature coefficient of resistance.

If two wires of different metals are used to form a junction, a potential is developed across the junction. This is because, when wires of different metals are joined together, electrons from the metal with the higher Fermi energy

See Semiconductor Devices.

11

level diffuse into the other metal until the levels become equal. This is the **Peltier effect**, and the potential difference thus created is the **contact potential**. It is dependent on the temperature of the junction since temperature affects the energy levels. If the dissimilar wires are connected at both ends to form a loop, the two contact potentials cancel, but, if the junctions are at different temperatures, a current circulates in the circuit. This effect is known as the **Seebeck effect**, after its discoverer in 1821, and is used in the thermocouple. Contact potential depends only on the temperature of the junction and not that of the interconnecting wires. A further metal can be connected in the loop without upsetting the emf as long as the two new junctions are at a common temperature. The contact potentials are very small and must be amplified for subsequent use.

Capacitance

Capacitive transducers rely on the basic relationship between the capacitance of two parallel plates and the distance between them. The capacitance is given by

$$C = \epsilon A / x$$

This expression assumes a perpendicular field with no fringing or edge effects.

where ϵ is the permitivity, A is the surface area of each plate and x is the plate separation.

By differentiating this expression with respect to the plate separation, x, we obtain $dC/dx = -\epsilon A/x^2$, and, by substitution for ϵA, we arrive at the expression $dC/C = -dx/x$. Thus a certain percentage change in the separation between two parallel plates produces a corresponding change in the capacitance. The relationship between capacitance and separation is non-linear, but for small changes, where dx is very small compared with x, the relationship between dC and dx is essentially linear.

Piezoelectrical transducers make use of the property of certain crystals to generate an electric charge when the crystal is deformed. Conversely, the crystal will deform when a charge is applied. Naturally occurring crystals such as quartz and rochelle salt exhibit the effect very strongly, and synthetic materials are now also widely used. The latter are of two main types: crystalline structures such as lithium sulphate and ammonium dihydrogen phosphate, and polarized ferroelectric ceramics such as barium titanate. The ferroelectric ceramics are polarized by heating to a temperature above the Curie point and allowing to cool slowly with a strong electric field maintained throughout the process. As the ceramic cools, there is a redistribution of molecular charges in a preferred direction dictated by the applied field. Subsequent distortion of the lattice generates a potential difference across the crystal faces. In order to make use of this, it is necessary to attach metal electrodes which effectively form a parallel-plate capacitor, giving a voltage

The Curie Point is the temperature at which the ferroelectric properties break down.

$$V = Q/C$$

where Q is the total charge and C is the capacitance.

The leakage resistance, through which the charge leaks away, is very large, being in excess of $10^{10}\ \Omega$, but the charge developed by a deformation inevitably decays, and this decay period is likely to be shortened considerably when the necessary connecting leads and sensing amplifier are attached. In order to limit the shunting effect of the external circuitry, a high-impedance buffer amplifier is normally used, and modern microelectronic techniques allow this to be built in to the transducer, very close to the sensor itself.

Electromagnetic tranducers

Electrical signals are generated when a conductor is moved through a magnetic field, or, conversely, if a magnetic field cuts the conductor. This is a statement of Faraday's Law, but it is usually quoted in the form attributed to Neumann:

$$E = -\mathrm{d}(N_\phi)/\mathrm{d}t$$

Faraday's law states that the magnitude of the induced emf in a circuit is directly proportional to the rate of change of the flux linkages.

where $\mathrm{d}(N_\phi)/\mathrm{d}t$ is the rate of change of the flux linkages in webers per second. This effect is used in the magnetic audio cartridge mentioned earlier. Movement of the stylus causes a flux pattern, which is effectively a change in flux linkages through the coil, and a voltage proportional to the stylus movement is generated across the coil. The same principle is used in magnetic proximity detectors, where it is necessary to detect moving objects such as, for example, the gear teeth on a revolving wheel.

Just as movement can be detected by the disturbance of a magnetic field, so movement can be created by the generation of a magnetic field. One of the most common requirements of a control system is for a means of mechanical movement, either linear or rotary. Almost all electromechanical actuators are based on the principle of magnetic attraction or repulsion: when a current flows through a solenoid coil, the magnetic field produced causes the soft iron core to move into the coil. This action can provide movement over short distances, and, typically, forces of 0.1 to 1 kg over distances of 5 mm are obtained. The loudspeaker is a variation of the idea so that, instead of the soft iron core moving, a speech coil, rigidly fixed to the cone, or diaphragm, moves over the magnetic pole piece. The pole piece is so arranged that the magnetic field is uniform, ensuring that the cone movement is proportional to the current through the coil.

Semiconductor transducers

The semiconductor materials found most commonly in modern electronic systems are silicon, gallium arsenide, germanium and cadmium sulphide. The extent to which any of these semiconductors conducts electricity is influenced by the presence of impurities within the crystal lattice. Very pure semiconductors are known as **intrinsic** semiconductors, those containing impurity atoms of elements such as antimony and indium are referred to as **extrinsic**. A single crystal of semiconductor material can be doped by the introduction of selected impurities in such a way that, although still electrically neutral, the crystal lattice acquires either an excess of electrons, becoming an n-type, or a reduction in the number of electrons, becoming a p-type material. A p–n interface forms a rectifying junction and will exhibit diode action. This is not the place to discuss the properties of semiconductors in great detail and we shall merely state that the diode characteristics can be described with a fair

13

This equation ignores such factors as the bulk resistance of the diode material. For a detailed discussion see *Semiconductor Devices*.

degree of accuracy by the equation

$$I = I_0[\exp(qV/kT) - 1]$$

where I_0 is the saturation, or leakage current, q is the charge on an electron, k is Boltzmann's constant, T is the absolute temperature and V is the applied voltage.

In the reverse direction, V is negative and as V increases the expression rapidly becomes

$$I = -I_0$$

In the forward direction of applied voltage, the diode current does not begin to grow until a 'cut-in' voltage has been reached, the size of the voltage depending on the work function of the material. For silicon this voltage is about 600 mV, and for germanium 300 mV. The leakage current, I_0, is highly temperature dependent and can be expressed as

$$I_0 \propto T^{3/2} \exp(-W/2kT)$$

where W is the work function.

For silicon, I_0 approximately doubles for every 7 °C rise in temperature, whereas for germanium I_0 doubles for every 10 °C rise. Though the leakage current in silicon shows a greater temperature dependence, the absolute value of current is much less. Typically, at 25 °C and with $V = -10$ V, I_0 for silicon is 25 nA, whereas for germanium under the same conditions I_0 is 10 μA. It is the leakage current which limits the range of temperatures over which diodes can be operated satisfactorily, the upper limit for silicon being about 130 °C, and for germanium about 75 °C. However, as we shall see in the next chapter, temperature sensors can be constructed to take advantage of the leakage current. Devices are available in integrated circuit form which can provide an output current directly proportional to absolute temperature.

Another method of temperature sensing is provided by the thermistor. This is a thermally sensitive semiconductor resistor formed from the oxides of various metals and packaged in a range of small beads, discs, rods, washers and probes. Semiconducting oxides of cobalt, copper, iron, manganese, nickel and titanium are commonly used, and the composition depends on the resistance value and temperature coefficient required of the thermistor. Because of their small size and low thermal inertia, changes with temperature can be much faster than with resistance temperature detectors. However, though the changes in resistance are larger, they are non-linear and are directly useful over a much narrower range, normally approximately -30 °C to approximately $+200$ °C.

In such cases, we can alternatively express the resistance as

$$R_T = A \exp(B/T)$$

Note that the temperatures here are in degrees Kelvin.

where R_T is the resistance at T K and B and A are constants for the material. A more convenient form of the expression is found by considering the difference in resistance at two temperatures, T_1 and T_2. Then

$$R_2/R_1 = \exp[B(1/T_2 - 1/T_1)]$$

B has typical values between 3000 K and 5000 K.

14

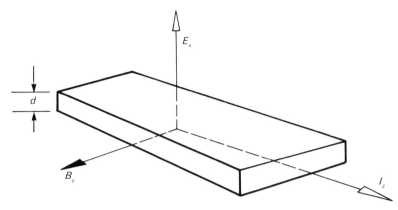

Fig. 1.11 The Hall effect

It is appropriate to include here another effect which is present to a limited extent in all conductors but which is very pronounced in semiconductors. This is the Hall effect which is utilized extensively in measurements related to magnetic fields, and is also used in non-contacting switches for keyboards and panel controls. The main component is a wafer of semiconductor material which is subjected to a magnetic field. When a current is passed through the wafer, normal to the magnetic field, a transverse voltage is developed across the wafer, which is proportional to the product of the magnetic flux density and the current. This can be shown quite simply by reference to Fig. 1.11. The current, I_z, results from the flow of charge carriers, and

$$I_z = nqvA$$

where n is the number of charge carriers per unit volume, q is the charge on an electron, v is the mean velocity of the charge carriers and A is the cross-sectional area of the wafer.

When the magnetic field, B_y is applied, the charge carriers experience a force causing the mobile carriers to migrate to one face of the wafer, leaving a residual opposing charge on the other face. The resulting electric field, E_x creates another force on the charge carriers which opposes that created by the magnetic field. At equilibrium, the force due to the electric field, $F = Eq$, equals the force due to the magnetic field, $F = Bqv$, giving $E = Bv$. The voltage resulting, if the wafer thickness is d, is

$$V = Ed$$

$$= Bvd$$

$$= BId/nqA$$

indicating that V is proportional to BI.

Commercial Hall effect magnetometers are very simple and robust, and can cope with field strengths from about 0.1 millitesla up to 1 tesla, which is the level of a strong permanent magnet.

Think of the current as flowing in an imaginary wire inside the wafer. Regardless of whether the current is electrons flowing in one direction or holes flowing in the opposite direction the current direction is the same, so the 'wire' will move in the direction indicated by Fleming's Right Hand Rule. For example, if the wafer in Fig. 1.11 is p-type, i.e. the current is hole current, then the upper face will become positively charged.

Photoelectrical effects

Photoelectrical transducers are those which either respond to the presence of

15

light, generating an electrical voltage or current, or generate light in response to the application of an electrical signal. The light can be in the visible range or, more frequently, in the longer wavelengths of the near infrared. Various light sensitive devices are available and can be grouped into three main types; phototransistor and photodiode (including light-emitting and laser diodes), photoconductive and photovoltaic.

The simplest optical detectors are photosemiconductors fabricated as diodes or transistors. All semiconductor junctions are sensitive to light and these detectors are similar to conventional devices but are packaged in transparent cases, so that the light can reach the junction. When radiation falls on the junction it creates hole-electron pairs in the depletion region, and, if correctly biased, a current flows in the external circuit. A photodiode will respond well only to a high light level. The usual sensitivity is up to about 1 A per watt of incident light, but most operating light levels reach only about one milliwatt, so the current level is normally low. A fast response of a few tens of nanoseconds is possible from a standard photodiode when operated in reverse bias, but for very fast response a PIN diode is preferred, giving switching times of less than a nanosecond. As the frequency of the incident light increases, the hole–electron pairs are generated closer to the surface of the material, and so further from the junction. This occurs because the light absorption coefficient of the material increases with frequency. There is, therefore, a limited range of frequencies over which an appreciable current is produced, and for most semiconductor devices this lies in the infrared region of the spectrum (Fig. 1.12).

As with all semiconductor devices, photodiodes have a temperature-dependent leakage current and this gives rise to a dark current in the diode, even when no light is present. It is this dark current which sets a limit to the ability to detect low light levels. For some specialized applications, where extremely low light levels are to be detected, photodiodes are cooled down to sub-zero temperatures using liquified gases.

PIN (p-intrinsic-n) diode is, at radio frequencies, an almost pure resistance whose value can be varied over a range from 1 to 10 000 ohms by a direct or low-frequency current.

Fig. 1.12 The visible and near infrared spectrum

A phototransistor relies on the same effect as the simple diode but has the current amplifying capability of the transistor built in. The emitter current is given by

$$I_E = (1 + h_{FE})I_p$$

where I_p is the photon-generated base current and h_{FE} is the dc current gain in the transistor. In order to give a high sensitivity to light a large collector-base junction area is used, and a high current gain. The current gain, h_{FE}, varies with bias levels and with temperature, and the performance of the phototransistor can easily be affected by time constants in the circuit in which it is operating. In general, the higher the circuit gain the slower the device responds to the light level changes. The features necessary to ensure a high sensitivity unfortunately also cause high dark current levels, since

$$I_{CEO}(\text{dark}) = h_{FE}I_{CBO}(\text{dark})$$

At 10 V reverse bias and 25 °C, a typical dark current is around 10 nA and increases with temperature in the usual way, so that, as a good approximation,

$$I_{CEO}(\text{dark}) \propto T^{3/2} \exp\left(-W/2kT\right)$$

where W is the work function of the material, k is Boltzmann's constant and T is the temperature.

A still higher current gain is given by the photo-darlington detector. The output current is $I_c = I_p h_{FE}$, where h_{FE} is in the range 10^3 to 10^5. We therefore obtain a good response even with very low light levels, though the dark current effects are also magnified and the speed of response falls to, typically, several microseconds.

In a forward-biased semiconductor diode, electron-hole pairs are created using the energy supplied by an electric field. When these charge carriers recombine under favourable circumstances light energy is emitted. In silicon and germanium, most of the recombination energy is dissipated as heat, but some semiconductors, such as gallium arsenide and gallium phosphide, are particularly efficient at producing light energy. The light from a light-emitting diode, LED, is incoherent but, by creating parallel reflecting surfaces on the semiconductor crystal, coherent, laser, light can be produced. This light is highly directional and can provide for efficient coupling into optical fibres. Laser diodes used in data communications operate in the infrared range at wavelengths of 800 to 900 nm.

Cadmium sulphide (CdS) and cadmium selenide (CdSe) cells are photoconductive devices, or light-dependent resistors, which respond to incident light by decreasing in resistance as the light intensity increases. The resistance is an inverse function of the majority charge carrier density, which increases as more light energy is provided. A typical visible light-dependent resistor will have a useful illuminated resistance range from about 100 Ω to about 2000 Ω with a dark resistance of several megohms. A CdS cell has a response time of several hundred milliseconds, though a CdSe cell is somewhat faster.

A third type of photoelectrical transducer is the photovoltaic cell, or solar cell, sometimes also referred to as a barrier photocell. This transducer actually converts the light energy to electrical energy and so needs no

Some phototransistors have an external connection to the base and the base current is then $(I_p + I_B)$. The sensitivity can be adjusted by varying I_B.

The photo-Darlington detector uses a compound transistor (Darlington pair) formed from two separate transistors giving an overall forward current gain of $h_{FE1} \times h_{FE2}$. See *Transistor Circuit Techniques*, Ritchie.

external power supply. The cell consists of a high-resistance photosensitive barrier layer between two conducting layers, and acts rather like a self-charging capacitor. When illuminated, the cell develops an output voltage of approximately 0.5 V, but the conversion efficiency is very low and a light energy of the order of 1 kW/cm^2 is required to sustain a current of a few tens of milliamps.

Silicon Technology

See *IEE Electronics and Power*, Feb. 1983, special feature on Measurement, Instrumentation and Transducers: also *IEEE Spectrum*, July 1990, 'Silicon Micromechanics'.

Recent years have seen the application of silicon integrated technology in the development of high-performance, low-cost sensors. As we have seen, silicon is a highly effective material for transducing many physical parameters including force, temperature and light. The present state of the technology already allows the high precision required in many sensing applications, and many further developments can be expected in this area. Many silicon transducers to date consist only of the basic sensing element, but in future we will expect more transducers to have the signal-conditioning circuitry fully integrated with the sensor. In due course the devices will also provide the output in a fully coded digital form.

Summary

This chapter has introduced the basic concepts underpinning many of the transducers used in process control. The forms transducers can take are endless, each application may demand a different design and an understanding of the underlying principles is therefore essential. From the basic definition of the transducer as a device which accepts energy from one system and delivers it, usually in a different form, to another system, the important general parameters have been indicated, and these must be taken into account when choosing and using transducers.

Review questions

1. Explain the difference between the terms 'transducer' and 'sensor'.
2. Define transducer sensitivity.
3. Describe a digital temperature sensor.
4. Describe three different types of pressure transducer.
5. Discuss the importance of repeatability in transducer performance.

Further reading

1. *Principles of Measurement Systems*, J P Bentley, Longman, 1983.
2. *Fundamentals of Modern Digital Systems*, B R Bannister and D G Whitehead, 2nd edn, Macmillan, 1987.
3. *Semiconductor Devices*, J J Sparkes, Van Nostrand, 1987.

4. *Digital Logic Techniques – Principles and Practice*, T J Stonham, 2nd edn, Van Nostrand, 1987.
5. *Sensor Review*, an international journal published quarterly by IFS (Publishers) Ltd.
6. Transistor Circuit Techniques discrete and integrated (2nd Edn) G. J. Ritchie.

Problems

1.1 A transducer with an internal resistance of $10\,\text{k}\Omega$ generates an output voltage which is sensed by a voltmeter having an internal resistance of $30\,\text{k}\Omega$. What percentage error is to be expected?

1.2 A negative temperature coefficient thermistor has a resistance given by $R_t = k \exp(B/t)$, where t is the temperature in degrees Kelvin. At $0\,^\circ\text{C}$ the resistance is $500\,\Omega$ and at $100\,^\circ\text{C}$ this has fallen to $40\,\Omega$. Find the resistance at $25\,^\circ\text{C}$.

1.3 By using the series/parallel arrangement shown it is possible to modify the resistance variation with temperature. What values should R_s and R_p have if the circuit is to present a resistance of $30\,\Omega$ at $100\,^\circ\text{C}$ and $65\,\Omega$ at $25\,^\circ\text{C}$?

1.4 A pressure transducer has an input in the range 0.0 to 5.0 Pa. Using the calibration results given in Table 1.1, estimate

(a) the maximum non-linearity as a percentage full-scale deflection;
(b) the maximum hysteresis as a percentage of full-scale deflection;
(c) the slope of the ideal characteristic.

Table 1.1

Pa	0.0	0.5	1.0	1.5	2.0	2.5	3.0	3.5	4.0	4.5	5.0
Output (mV) increasing	0.0	7.0	14.5	20.0	25.0	27.0	31.5	35.5	39.0	42.0	45.0
Output (mV) decreasing	0.0	3.0	6.5	9.5	13.5	17.5	22.0	26.5	31.0	37.0	45.0

1.5 A wirewound translational potentiometer consists of a cylinder 60 mm long and 10 mm diameter. Around its circumference are wound 500 turns of resistance wire, evenly spaced. The diameter of the wire is 0.1 mm and its resistivity is $50 \times 10^{-8}\,\Omega\text{m}$. If the wiper can move from one turn to the next without touching two turns at once calculate the resolution as a percentage of full-scale deflection. If the potentiometer is loaded with $1\,\text{k}\Omega$ across its output terminals, what is the maximum error?

2
Sensors, actuators and displays

In this chapter we look at some of the numerous types of transducer which are commonly available. As a first step we have already differentiated between input and output transducers and we now further subdivide, but still in broad terms. Input transducers, or sensors, can be grouped according to the physical quantities that are to be detected and, in many cases, quantified. The main categories are:

1. mechanical sensors for force, pressure, position, proximity, displacement, velocity, acceleration, vibration and shock;
2. sensors for temperature;
3. radiation detection;
4. chemical activity.

Output transducers are grouped according to their purpose:

1. actuators for control;
2. displays for information.

Mechanical sensing

When one considers the demand for weighing and pressure sensing in modern equipment, it is not surprising that force sensors are probably the most common of transducers, and almost all sensors designed to detect and measure force rely upon the transformation of the force into the deformation of an elastic medium. For example, the application of a longitudinal force to a steel rod will result in an extension of that rod by an amount

$$\mathrm{d}x = EXF/A$$

where E is Young's modulus for the rod material, F/A is the applied force per unit area, i.e. the stress, and X is the original length of the rod.

Provided that the force applied is not so great that the rod is permanently deformed, then the above expression applies and the extension, $\mathrm{d}x$, bears a linear relationship to the applied force. The variable to be measured now becomes a more tangible physical displacement, or strain, where strain is defined as the change in length per unit length.

Thus

$$\mathrm{d}x/X = EF/A$$

or

$$\text{strain} = \text{Young's modulus} \times \text{stress}$$

When a material is subjected to a linear tensile or compressive force, the change in length produced is related to the force by Young's modulus of elasticity, which is defined as $E = \text{stress/strain}$, where stress is in N/m^2 and strain is the change in length per unit length.

The strain gauge

First developed in the USA in the late 1930s, the strain gauge is now the most widely used device for the measurement of force. It consists of a resistive material, initially a length of fine wire but now usually a metal foil, only a few microns in thickness, made by etching a grid pattern in copper-nickel or chrome-nickel (Fig. 2.1). The foil can be produced in many sizes

Micrometres are also known as 'microns'.

direction of stress

metal film

polyester backing

Fig. 2.1 The strain gauge; these are commonly fabricated on a polyester film which is then glued to the specimen under test and shapes to suit the requirements of the transducer. It also has well-defined thermal properties which permit accurate operation over a wide temperature range. When under strain, the resistance of the foil changes. In use, the gauge is bonded to a suitable carrier which is to be subjected to the force. The gauge is mounted so that the long lengths of the conductor are aligned in the direction of the force to be measured. This force, or more precisely the stress, gives rise to a strain in the carrier and its bonded foil, and the subsequent change in resistance of the metal foil is measured electrically. Gauges are available with automatic temperature compensation designed to match the expansion with temperature of the most commonly used construction metals, such as aluminium and stainless steel.

The resistance of a conductor is given by

$$R = \rho L / A$$

where ρ is the resistivity, L is the length and A is the cross-sectional area. By its nature, the change in resistance is proportional to the average strain over the foil and, for accurate measurements, it is important that the shape of the carrier should ensure that the stress in the region of the gauge correctly reflects the forces to be measured. Now, the gauge material undergoes a change in both length and cross-sectional area when strained, and the resistivity of the material also alters as a result of the piezoresistive effect. The ratio of the percentage change in resistance to the percentage change in length is known as the **gauge factor**, K, and is a useful parameter in quantifying strain gauges. In its complete form, the gauge factor is given as

An in-depth treatment of gauge factors is to be found in *Measurement Systems*.

$$K = \frac{\mathrm{d}R/R}{\mathrm{d}L/L} = 1 + 2v + \frac{\mathrm{d}\rho/\rho}{\mathrm{d}L/L}$$

where the term $2v$ takes account of the change in cross-sectional area, and the term

v is Poisson's ratio which for most metals is approximately 0.3.

$$\frac{\mathrm{d}\rho/\rho}{\mathrm{d}L/L}$$

$$\frac{\mathrm{d}\rho/\rho}{\mathrm{d}L/L}$$

is usually about 0.4.

relates to the piezoresistive effect.

Typically, gauge factors lie between 2 and 4 for the commonly used metal foil gauges, and their resistance between 100 and 1000 Ω.

In practice, the change in resistance is very small and strain gauges are almost invariably used in a bridge configuration where the change in resistance is converted to a voltage variation for subsequent amplification. Consider, for example, a balanced Wheatstone bridge as shown in Fig. 2.2.

Fig. 2.2 A strain gauge bridge

All resistances are of equal value initially, but one of them, R^*, is a strain gauge whose resistance increases by dR. The out-of-balance voltage, V_o, is given by

$$V_o = V_s dR / (4R + 2dR)$$

or, rearranging,

$$V_o = \frac{V_s}{2R} \frac{dR}{(2 + dR/R)}$$

This expression shows that the response to a change in resistance is a non-linear change in voltage. However, as the fractional change in resistance is very small, to a good approximation we have the linear relationship

$$V_o = \frac{V_s dR}{4R}$$

Since

$$K = \frac{dR/R}{dL/L},$$

we obtain the expression

$$V_o = \frac{K}{4} V_s \frac{dL}{L}$$

showing that the voltage is proportional to strain. This is an important relationship because it shows that, if the excitation voltage, V_s, and the gauge factor, K, are known, it is only necessary to measure the out-of-balance voltage, V_o to obtain the level of strain.

A single active 120 Ω strain gauge of gauge factor 2.0, wired into a standard bridge circuit, is mounted onto a steel specimen. Calculate the minimum stress that can be detected if the maximum sensitivity of the monitoring equipment is 10 μV. The bridge supply is 8 V and Young's modulus for steel is 2.1×10^5 MN/m. For the bridge shown in Fig. 2.2,

$$V_o = (\text{strain}) K V_s / 4$$

Young's modulus, $E = \text{stress}/\text{strain}$ so

$$\text{stress} = 4vE/KV_s$$

and the minimum stress is

$$\frac{4 \times 10 \times 2.1 \times 10^5}{2.0 \times 8} = 0.525 \text{ MN/m}$$

Worked example 2.1

It should be remembered that in a practical system sensitivity is limited by the signal-to-noise ratio. All electronic equipment generates noise signals, and the limit in sensitivity occurs when the signal cannot be detected above this noise level.

The sensitivity of the bridge can be doubled if two gauges are used, connected in opposing arms of the bridge. In practice, most strain-gauge bridges use four gauges; two opposing gauges are mounted so that their length is in the direction of the stress, and the other two opposing gauges are mounted normal to the stress. This construction causes resistance change due to temperature effects to cancel out, but note that it cannot cancel out the effects of strain in the carrier caused by temperature variations.

The construction of the carrier determines the nature of the transducer. A suitable primary transducer is constructed to carry the foil gauge. This primary transducer serves to convert the quantity that is to be measured into strain and the foil gauge then converts the mechanical strain into a change in resistance. Any physical quantity, capable of subjecting the carrier to mechanical stress, can be measured in this way. For example, acceleration can be measured if a mass is supported by a thin flexure beam, with the foil gauge bonded to the beam in such a way that when the mass undergoes acceleration, causing the beam to flex, the foil is subjected to strain (Fig. 2.3). Mounting the structure in a silicone-oil-filled enclosure provides suitable damping. Pressure can be measured by mounting strain gauges on a flexible diaphragm which serves to separate the medium being monitored from either a vacuum (absolute pressure measurement) (Fig. 2.4), or from atmospheric pressure (relative pressure measurement). Commercial devices are often provided with an optional vent.

Semiconductor strain gauges utilize the piezoresistive effect existing in semiconductor materials. The gauges are of two types: the bar gauge, formed from a bar of doped silicon cemented to the carrier which is to be stressed, in a similar manner to the foil gauge, and the diffused gauge, in which the silicon substrate itself acts as the carrier. In the latter case the substrate, typically with an area of less than 5 mm, is etched into the form of a diaphragm, and the transducer is created by diffusing n-type or p-type material at appropriate points on the diaphragm to give a four-element bridge

Fig. 2.3 Measurement of acceleration with two active and two temperature-compensating gauges

Fig. 2.4 Measurement of pressure

Fig. 2.5 Thin-diaphragm silicon pressure sensor

as in Fig. 2.5. Gauges made from p-type material increase their resistance with increasing strain, whereas n-type gauges decrease in resistance.

The main feature of the semiconductor strain gauge is the very high gauge factor, though this tends not to be constant with strain. Also, the change in gauge resistance is far more temperature dependent than is the case with metal gauges. In spite of these limitations, however, the advantages of silicon technology, with its suitability for high-volume, batch-mode production, make the diffused gauge an important modern device.

The strain-gauge transducer converts displacement (strain) into a change in resistance, and the magnitude of this displacement is generally quite small. Typically, the displacement occurring when a working stress is applied to a load cell is only a thousandth of an inch or so (about $25\,\mu\mathrm{m}$). Another commonly used device for converting small displacements into electrically measurable quantities is the capacitance transducer. This device demands more complicated electronic support circuitry than the strain gauge but the mechanical simplicity of construction and the high sensitivity make it popular in many applications. The essential constructional details of a capacitance

Gauge functions for p-type silicon vary between +100 and +175, and for n-type silicon between −100 and −140.

The name 'load cell' refers to a transducer which produces an output that is proportional to an applied force.

25

Thin flexible conducting diaphragm

Reference pressure

Electrode

Connections

Fig. 2.6 Cross-section of a capacitive pressure transducer

Fig. 2.7 Capacitance bridge

At balance $C_t = C_1 R_2 / R_1$.

pressure transducer are shown in cross-section in Fig. 2.6. A change in pressure produces a displacement of the diaphragm, which alters the capacitance between the centre electrode and the outer casing. To measure this capacitance, a capacitance bridge is used (Fig. 2.7). The bridge is initially balanced for minimum V_o, and any change in the magnitude of C_t unbalances the bridge. Noting the value of R_2 necessary to rebalance the bridge allows the change in C_t to be calculated. The sensitivity of the transducer increases as the plate separation is decreased.

Worked example 2.2

See p. 12.

A parallel-plate variable airgap capacitance transducer has a plate area of 100 mm². If the separation is 0.2 mm, calculate the sensitivity in pF/mm.
The capacitance of a parallel-plate capacitor, neglecting edge effects, is

$$C = \epsilon_0 \epsilon_r A / x$$

where ϵ_0 is the permitivity of free space (8.854×10^{-12} F/m), ϵ_r is the relative permitivity of air (1.0005, i.e. ≈ 1.0), A is the area of the plates and x is the separation.
The sensitivity is defined in Chapter 1 as

$$S = dC/dx$$

$$= -\epsilon_o \epsilon_r A/x^2$$
$$= \frac{8.854 \times 10^{-12} \times 10^{-4}}{(0.2 \times 10^{-4})^2} = 221 \times 10^{-10} \text{ F/m}$$
$$= 22 \text{ pF/mm}$$

Perhaps the most common pressure capacitance transducer is the capacitance microphone, in which the varying capacitance caused by the changes in air pressure in the sound wave give rise to small charge and discharge currents which, in turn, are monitored as changes in voltage across R (Fig. 2.8). Typically R has a value around a megohm. The output voltage is only a few millivolts and the necessary amplifier must have a very high input impedance to avoid excessive loading of the circuit. FET amplifiers are normally used, and most commercial 'condenser' microphones are supplied with the amplifier built in.

To measure larger displacements, where the inherently non-linear inverse relationship of the characteristics cannot be ignored, the circuit shown in Fig. 2.9 can be used. The operational amplifier has a very high gain and high input impedance, so that both current and voltage at the amplifier input are negligible. From the circuit

Op-amps are considered in detail in the next chapter.

$$V_o - V_i = \frac{1}{C_t} \int i_t \, dt$$

and

$$V_s - V_i = \frac{1}{C_1} \int i_i \, dt$$

Since V_i approximates to zero, then

Fig. 2.8 Equivalent circuit of capacitance microphone

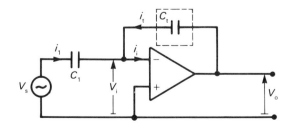

Fig. 2.9 Use of an amplifier where displacements are large

$$V_o = \frac{1}{C_t} \int i_t \, dt \quad \text{and} \quad V_s = \frac{1}{C_1} \int i_i \, dt$$

but the amplifier input current, i_i, also approximates to zero so $i_i = -i_t$ giving

$$V_s = -\frac{1}{C_1} \int i_t \, dt \quad \text{and} \quad V_o = -V_s C_1 / C_t$$

However, the capacitance, $C_t = \epsilon A / x$, where $\epsilon = \epsilon_o \epsilon_t$, so

$$V_o = -V_s C_1 x / \epsilon A$$

Assuming a constant source voltage, V_s, we see that V_o is directly proportional to the diaphragm displacement, x.

Another popular device used in converting displacement into electrical signals, is the linear variable differential transformer (LVDT). Here, motion of a magnetic core varies the mutual inductance of two secondary coils relative to a primary coil. LVDTs can be obtained which measure displacements of the same order as strain gauges, but they are capable of much greater movement and ranges up to ±3 inches (75 mm) are not untypical. The LVDT is simple in construction, and is basically a transformer with the primary coil sandwiched between two secondary coils as shown in Fig. 2.10. The magnetic core is free to move along its axis. If an ac signal, typically 3 to 15 V, is applied to the primary, a null position can be found by moving the core so that the voltages induced in each secondary coil exactly cancel out. At this point the mutual inductances between each secondary coil and the primary coil are equal. In practice, however, the null point is difficult to achieve, due to stray capacitance between the primary and secondary windings. Usually a null can be obtained to within 1% of the full-scale output voltage and, if this is not adequate, a judicious blend of ground shielding and electronic balance will improve matters. The output voltage of the LVDT is a linear function of the core displacement, within a specified range of movement, and it can be shown that the output voltage, e_s, is given by

$$e_s = e_{s_1} - e_{s_2} = (M_1 - M_2) \frac{di_p}{dt}$$

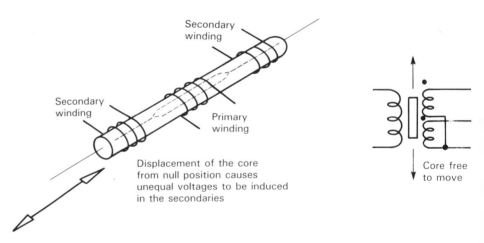

Fig. 2.10 The linear variable differential transformer (LVDT)

where M_1 and M_2 are the mutual inductances of the secondary windings and i_p is the instantaneous primary current.

The direction of core displacement is indicated by the phase of the output signal, and the linearity is determined by the range over which the mutual inductance varies linearly with core position.

The LVDT can offer serveral distinct advantages over many other types of displacement transducer. There is no mechanical wear, resulting in an extended life; complete electrical isolation exists between input and output, allowing easy interfacing to electronic equipment; and damage caused by over-ranging is virtually impossible. In addition, LVDTs generally have a high level of output, lying typically between 1 and 5 mV per volt excitation for each thousandth of an inch displacement. Naturally, since we are dealing with inductive devices, in which impedance varies with frequency, the output must be qouted for some specific working frequency, which can be as high as 20 kHz. There are some 'dc' LVDTs which take advantage of modern integration techniques, and include an oscillator, driver and rectifier plus signal conditioning circuits all in the single housing. Interfacing of such units is particularly straightforward.

A common application of displacement transducers, such as the strain gauge and the LVDT, is in load cells used in weighing equipment ranging from fill-weighers, weighing up to several kilograms, to weighbridges for checking the weight of goods vehicles. Figure 2.11 shows a typical arrangement using strain gauges in an axle weighing application. Four transducers support a steel platform, and a vertical force causes a strain or displacement to occur in the gauges mounted circumferentially on a steel spindle within each transducer.

An interesting non-contact technique for the measurement of depth uses laser beam triangulation. The principle is simple and is shown diagrammatically in Fig. 2.12. A semiconductor laser diode emits pulses of light which create an illuminated spot on the surface to be measured. A small proportion of the light scattered by the surface is picked up by a camera unit and focused on to a light-sensitive detector. The detector is a linear device consisting of a sequence of light-sensitive cells, often 256, from which it is possible to determine at what point along its length the light impinged. Any change in the distance to the measured surface results in a change in the position of the

Fig. 2.11 Use of strain gauges in a weighbridge

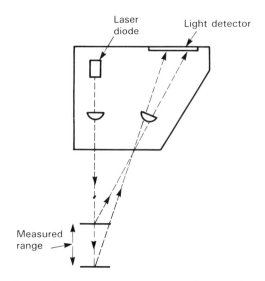

Fig. 2.12 Depth measurement by laser beam triangulation

focused spot on the detector. Typically, gauges can be obtained with measurement ranges from 8 mm to 512 mm, with an accuracy of 0.1% of the measurement range, and are increasingly being used in remote sensing applications where contact with the surface is not feasible. The technique can be used to provide data on depth, profile, thickness, vibration, or indeed any parameter which changes with the position of the surface.

The synchro

A **synchro** is a form of rotating transformer resembling a small ac motor, and is widely used as an element of measuring and control systems involving rotatable shafts. The primary coil is wound on the rotor and couples magnetically to the secondary coils on the stator. Connection to the external terminals is via sliprings.

The forerunners of today's synchros were called 'magslips' – a contraction of 'magnetic sliprings'.

The synchro commonly has three secondary windings arranged 120° apart to give varying voltage ratios as the primary winding is rotated (Fig. 2.13). The rotor energizing voltage is an ac reference voltage at 50 Hz (or in some cases 400 Hz) and this induces voltages across each stator winding which are dependent on the instantaneous angle between the rotor and stator coil axes. The voltage across any pair of stator terminals is, therefore, the sum or difference, depending on the phase, of the individual stator winding voltages.

If at some instant the synchro shaft angle is θ and the rotor reference voltage is $V \sin \omega t$, the stator terminal voltages are

$$S_{1-3} = V \sin \omega t \sin \theta$$

$$S_{3-2} = V \sin \omega t \sin (\theta + 120°)$$

$$S_{2-1} = V \sin \omega t \sin (\theta + 240°)$$

Synchro windings

Fig. 2.13 Synchro windings

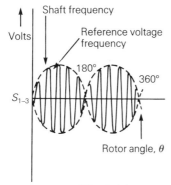

Fig. 2.14

If the rotor rotates continuously at a constant rate each stator terminal voltage appears as a sinewave at the reference voltage frequency, amplitude modulated at the shaft frequency. (Fig. 2.14)

These are known as the **synchro format voltages**.

In some cases the angular information from one synchro must be directly coupled to the shaft of another synchro and a pair of torque synchros, consisting of transmitter and receiver, are used. These synchros are able to handle the necessary power levels and form a **torque chain**. In general, however, synchros used in measurement and control applications, which rely on accurate angular information in servo systems, are not required to handle high torque levels and can operate at higher impedance levels.

Control synchros also operate in pairs to form a control chain, but consist of a control transmitter synchro and a control transformer synchro. The latter is similar to a torque receiver synchro but has its rotor winding aligned 90° out of phase, relative to the transmitter. As before, the voltages generated at the stator terminals of the control transmitter are governed by the relative angle of the rotor and stator windings.

If the control transmitter synchro and its control transformer synchro are connected together, as in Fig. 2.15, the voltage appearing at the transformer rotor terminals is directly proportional to the transmitter rotor angle. When the two shaft angles are identical a null voltage appears at the transformer rotor terminals and, as the alignment deviates, an increasing voltage is produced with a phase relative to the reference voltage which indicates the

Higher impedance means lower power.

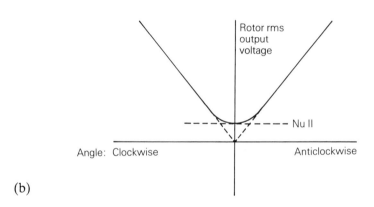

Fig. 2.15 (a) Synchro control chain. (b) Transfer rotor terminal voltage

direction of deviation. Ideally the null voltage is zero but in practice there is a residual voltage of a few tens of millivolts caused by unbalanced voltages at fundamental frequency and harmonic voltages.

A typical use of synchros in a shaft position control servo system is shown in Fig. 2.16. This commonly occurs in applications such as the control of an altitude radar system or the duplication of master compass information on-board ship, and, in fact, any equipment in which the angular position of one shaft is to be controlled by the position of another shaft. If the input shaft is set to a relative angle, θ_1, and the output shaft is at some other angle, θ_2, an error voltage proportional to $|\theta_1 - \theta_2|$ is produced on the transformer rotor winding and this is used to drive the servo motor in a direction such that the error is reduced to a minimum.

Resolvers are often called 'synchro resolvers'.

A **resolver** is a variant of the synchro and, in its simplest form, uses two stator windings displaced by 90° rather than the three displaced by 120° (Fig. 2.17). If the rotor reference voltage is $V_{ref} = V \sin \omega t$ and the shaft angle is θ, then

Fig. 2.16 Synchro control chain in a servo system

Fig. 2.17 The simple resolver

$$S_{1-3} = V \sin \omega t \sin \theta$$

and

$$S_{4-2} = V \sin \omega t \cos \theta$$

These are the 'resolver format voltages'.

These resolvers can be used in a similar way to synchros to transmit data in a control chain, using a resolver transmitter at one end and a resolver control transformer at the other. In general, however, a more complex form of resolver is used in which two rotor windings (at 90° to each other) are provided, as well as the two stator windings. One of the main advantages of this arrangement is that the device can be used as a **computing resolver** to perform polar-to-rectangular coordinate conversion or to deal with axis transformation. Both of these are often required in robot control and radar and guidance systems.

In polar-to-rectangular coordinate conversion we represent the polar co-ordinates of a point by a voltage $S_{1-3} = V \sin \omega t$, applied to one of the stator

33

windings, and the angle θ of the resolver shaft. The other stator winding is short-circuited. The resulting rotor voltages are

$$R_{1-3} = V \sin \omega t \sin \theta$$

and

$$R_{4-2} = V \sin \omega t \cos \theta$$

which represent the equivalent rectangular coordinates directly.

Similarly, when a voltage, V_1, is used to excite one of the stator windings and a voltage, V_2, excites the other stator winding, the rotor terminal voltages for a shaft angle of θ are

$$R_{1-3} = V_1 \cos \theta + V_2 \sin \theta$$

and

$$R_{4-2} = V_2 \cos \theta - V_1 \sin \theta$$

If voltages V_1 and V_2 respectively represent the rectangular coordinates of a point x_1, y_1 on a given set of axes, R_{1-3} and R_{4-2} represent the coordinates of the same point, x_2, y_2, on a second set of axes at an angle θ to the first set.

Worked example 2.3

Show that the voltages produced by the computing resolver accurately represent the new rectangular coordinates when the axes are rotated by an angle θ.

The origin of the axes is at point O on Fig. 2.18, and the point P then has coordinates x_1, y_1 on the first set of axes. When the axes are rotated through an angle θ, the point P has new coordinates x_2, y_2. Thus, $OA = x_1$, $AP = y_1$, $OB = x_2$, and $BP = y_2$. An additional construction draws a line through D parallel to the new y axis.

From Fig. 2.18 we see that

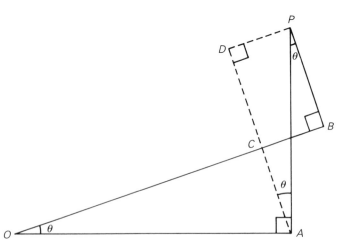

Fig. 2.18 Coordinate transformation

$$x_2 = OB = OC + CB$$
$$= x_1 \cos \theta + y_1 \sin \theta$$

Since *DP* = *CB*.

Also

$$y_2 = AD - AC$$
$$= y_1 \cos \theta - x_1 \cos \theta$$

Temperature sensing

Temperature-sensitive transducers are available in five main types. These differ in the ranges of temperature to which they react, and in sensitivity and stability. They also vary considerably in cost.

The types are:

Resistance temperature detectors (RTD)
Thermistors
Thermocouples
Semiconductor
Pyrometers

The **resistance temperature detector** (RTD) is essentially a length of wire wound on a bobbin and housed in a protective sleeve, and its operation is dependent on the variation of resistance with temperature of the wire. Platinum wire is normally used because of its stability over a wide temperature range, and its linear resistance characteristics. Nickel and copper are also used in less demanding applications. All these metals have positive primary temperature coefficients of resistance. The sensitivity of these devices, dR/dT, is low and the thermal inertia (the inability to respond rapidly to temperature changes) is rather high because of the construction. Also they are susceptible to shock and vibration.

A platinum resistance temperature detector is used to measure temperatures between 0 °C and 200 °C. The temperature coefficients are

Worked example 2.4

$$\alpha_1 = 3.96 \times 10^{-3} \, °C^{-1}, \, \alpha_2 = -5.85 \times 10^{-7} \, °C^{-2}.$$

If $R_0 = 100 \, \Omega$, find the sensitivity of the sensor at the two extremes of the temperature range. Find also the resistance at 100 °C and 200 °C.

The non-linear coefficients are usually small, and in this case we can ignore coefficients above α_2. Thus the general expression, from Chapter 1,

$$R_T = R_0(1 + \alpha_1 T + \alpha_2 T^2 + \alpha_3 T^3 + \ldots)$$

reduces to $R_T = R_0(1 + \alpha_1 T + \alpha_2 T^2)$. The sensitivity is then,

Note that *T* is in °C.

$$dR/dT = R_0(\alpha_1 + 2\alpha_2 T)$$

which evaluates to 0.396 at 0 °C and marginally less at 200 °C. The resistance at 100 °C is

$$R_{100} = 100(1 + 3.96 \times 10^{-3} \times 10^2 - 5.85 \times 10^{-7} \times 10^4)$$

$$= 139.0 \ \Omega$$

Similarly,

$$R_{200} = 176.9 \ \Omega$$

In use the sensor forms one limb of a standard bridge, but a complication arises from the fact that this sensor limb has to be physically separated from the rest of the bridge. The resistance of the sensor is only about 100 Ω for a platinum sensor, and much less for copper. This means that any contact resistance, or resistance in the leads connecting the sensor to the bridge, can be large enough to affect the accuracy of any measurement, and compensating techniques must be employed. One common method is to include a third lead from the detector (Fig. 2.19). Under this arrangement, known as the Siemens three-lead method, the resistance of lead L_1 is in one arm of the bridge, with the detector, R_t, and the resistance of lead L_2 in the adjacent arm with R. At balance, the current through lead L_3 is zero and the currents through L_1 and L_2 are equal, giving identical volt drops if the lead resistances can be made equal. The volt drops then cancel one another. The bridge has maximum sensitivity to changes in R_t when all four arms of the bridge have equal resistance. With a 10 V supply, as shown, the output, V_o, of the bridge is of the order of 1 mV/°C.

Modern forms of the resistance temperature detector include thin-film devices in which the resistance element is laid down as a zigzag metallic track a few microns thick on a ceramic substrate. Precise control of the resistance is achieved by laser trimming of the platinum track. The large reduction in size which this construction allows gives a much lower thermal inertia and a good sensitivity.

Thermistors provide a cheap and convenient way to measure temperature and are widely used in control systems where high accuracy is not required. They are essentially non-linear in operation, as we have seen in Chapter 1, but microprocessor-based systems can now be used to relieve the limitations this causes, allowing thermistors to be used over a much wider range of temperatures than hitherto. Knowing the equation constants, accurate values can be obtained by calculation, or by use of a look-up table. Different types of thermistor are available with either positive or negative temperature coefficients. The majority of thermistors used in sensing and measuring

Fig. 2.19 Three-lead compensation

(b)

(a)

Fig. 2.20 Simple temperature-measuring circuits

temperatures have a negative coefficient of about 4%/°C. Resistance values at 25 °C normally range from about 100 Ω to about 25 kΩ, though resistances greater than 500 kΩ are available. A simple temperature-measuring circuit (Fig. 2.20a) produces a varying voltage which is applied to an amplifier. Alternatively, the thermistor can form one arm of a bridge, as in Fig. 2.20b. The bridge component values determine the output voltage range and are, in general, chosen to give good linearity with reasonable sensitivity. The dissipation in the thermistor must also be restricted or its self-generated heat will affect its resistance values.

A thermistor with constants $A = 0.0085$ and $B = 3950$ K is used in a bridge, as in Fig. 2.20b, with the following component values: $R_1 = 230$ Ω; $R_2 = 1.4$ kΩ; $R_3 = 2.7$ kΩ; $V_i = 2.5$ V. Find the output voltage range as the temperature varies from 0 °C to 80 °C.

Worked example 2.5

See p. 10.

From the general bridge equation

$$V_o = V_i \left[\frac{1}{1 + R_t/R_3} - \frac{1}{1 + R_2/R_1} \right]$$

Thus the output voltage is governed by the input voltage, V_i, the values of R_t and R_3, and the ratio of R_2/R_1 (not their absolute values). If $\overline{R}_t$ and $\underline{R}_t$ are the maximum and minimum values of R_t respectively,

$$\underline{V}_o = V_i \left[\frac{1}{1 + \overline{R}_t/R_3} - \frac{1}{1 + R_2/R_1} \right]$$

and

$$\overline{V}_o = V_i \left[\frac{1}{1 + \underline{R}_t/R_3} - \frac{1}{1 + R_2/R_1} \right]$$

The given values indicate that $R_t = 0.0085 \exp(3950/T)$, and since the temperature range is 273 K to 353 K, the corresponding resistance values are $R_{273} = 0.0085 \exp(3950/273) = 16.337$ kΩ and $R_{353} = 615.3$ Ω. Thus,

$$\underline{V}_o = 2.5 \left[\frac{1}{1 + 16.337 \times 10^3/2.7 \times 10^3} - \frac{1}{1 + 1400/230} \right]$$

$$= 0.0025 \text{ V}$$

and

$$\overline{V}_0 = 2.5 \left[\frac{1}{1 + 615.3/2.7 \times 10^3} - \frac{1}{1 + 1400/230} \right]$$

$$= 1.683 \text{ V}$$

See Chapter 1.

Unlike resistance temperature detectors and thermistors, a **thermocouple** is a passive device which responds to temperature differences. Since it makes use of the Seebeck effect at a junction of two dissimilar metals, the thermocouple can be very small. Most thermocouples are made in the form of a probe, both for convenience in use and to protect the measurement junction. A remote junction is also required. The thermal emf generated by the thermocouple varies with the temperature difference between the junctions, and Fig. 2.21 shows the thermal emf developed by a typical iron–copper thermocouple with the reference junction held at 0 °C. The emf rises initially to a maximum at the neutral temperature, and then falls as the temperature continues to increase, eventually reversing in sign at the inversion temperature.

We saw in Chapter 1 that the addition of other leads to the thermocouple loop does not upset the effect, as long as the additional junctions are all at the same temperature. We can, therefore, use a voltmeter, or other suitable instrument, in order to measure the voltages. We can alternatively introduce

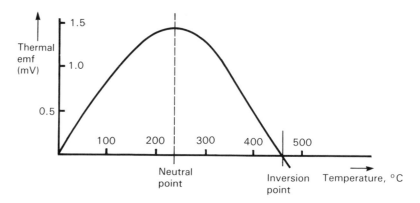

Fig. 2.21 Variation of thermal emf with temperature

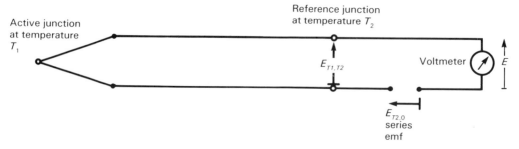

Fig. 2.22 Reference junction compensation

an emf, such as would be generated by an additional junction, in order to overcome a difficulty, which often occurs in practice, where it is inconvenient to maintain the reference junction at a constant 0 °C (or whatever other value is required). If we work at a convenient temperature of T_2 °C at the reference junction, we introduce a series emf, $E_{T_2,0}$, which corresponds to the potential which would be generated if one junction were at T_2 °C and the other at 0 °C. By summing the voltages around the loop formed by the reference junction, the voltmeter and the series emf (Fig. 2.22), we see that the emf at the voltmeter is then

$$E = E_{T_1,T_2} + E_{T_2,0} = E_{T_1,0},$$

and the net result is that an emf E appears at the voltmeter as though the reference junction were at 0 °C. This automatic reference junction compensation emf is normally provided by a special bridge circuit in which a resistance temperature detector is used.

Thermocouples have several advantages over other temperature sensors, being very small, low-cost, rugged devices, with wide temperature range and a low thermal inertia, but the sensitivity and output are very low, and a reference temperature is needed. Typical sensitivities range from 5 to 80 μV/°C.

The semiconductor diode equation, given in Chapter 1, indicates that a diode or transistor can be used as a temperature-sensing device. With a constant current source, I, and very small leakage current, the diode equation reduces to

$I = I_0[\exp(qV/kT) - 1]$.

$$V = \frac{KT}{q} \ln \frac{I}{I_o}$$

so that the junction voltage is proportional to temperature, and reduces by approximately 2.2 mV/°C. Many sensors based on this principle are commercially available. The National Semiconductor LM3911, for example, has a built-in amplifier and operates from a single supply to give an output of 10 mV/K.

A simple diode temperature detector can be used to control the frequency of oscillation of an oscillator circuit so that the frequency is directly proportional to the temperature. Alternatively, the diodes can be used in matched pairs in a bridge circuit, but the advantage of the low cost of the diode is then lost.

This is dealt with in Chapter 3.

In practice, the leakage current, I_o, can vary somewhat erratically since it is affected by carrier diffusion and recombination rates, and surface leakage effects, all of which are temperature dependent. As a first approximation

$$I_o \propto T^{3/2} \exp(-W/2kT)$$

where W is the work function of the material.

For silicon, the leakage current is low but doubles every 7 °C rise in temperature. For low-level signal diodes leakage current is measured in pA, but with silicon power rectifiers a typical leakage current of 25 nA at 25 °C has reached 6.5 mA at 150 °C, and it is this that limits the useful range of these devices.

More complex semiconductor circuits are now cheaply available in the form

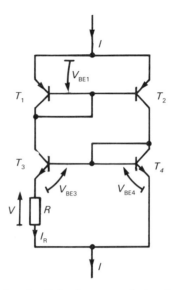

Fig. 2.23 Simplified circuit of the integrated circuit temperature sensor

of two-terminal transducers, housed in various standard transistor cases, which generate an output current, of a few microamps, directly proportional to temperature. The AD590, for example, is a high-impedance, constant-current source which provides a current of $1\,\mu A/K$ with a supply voltage between 4 and 30 V dc. Laser trimming of the internal thin-film resistors is used to calibrate the device so that its output at $25\,°C$ $(298.2\,K)$ is $298.2\,\mu A \pm 2.5\,\mu A$. The internal circuitry makes use of the current mirror technique, and a simplified form of the circuit is given in Fig. 2.23. Transistors T_1 and T_2 have the same base-emitter voltage, V_{BE1}, so the current I splits equally between them. Although shown as a single transistor, T_3 is, in fact, eight transistors, identical to T_4, in parallel, so that the current density in T_4 is eight times the density in any one of the T_3 transistors. The voltage, V, across the resistance R is the difference between two base-emitter voltages since, by Kirchhoff's voltage drop equation,

$$V = V_{BE4} - V_{BE3}$$

Thus

$$V = \frac{kT}{q} \ln\left(I_4/I_0\right) - \frac{kT}{q} \ln\left(I_3/I_0\right)$$

$$= \frac{kT}{q} \ln\left(I_4/I_3\right)$$

But the current in T_4 is eight times that in T_3 so

$$V = kT/q\,(\ln 8)$$

The voltage across R is thus directly proportional to absolute temperature, as must be the current through R, and also, therefore, the total current I. The total current,

$$I = 2I_R = \frac{2V}{R} = \frac{2kT}{qR} \ln 8$$

and by adjusting R to 358 Ω the ratio I/T can be made to be 1 μA/K. This type of device is particularly attractive since it requires no complex support circuitry for compensation or detection and, being a current source, it can be used in remote applications where contact and line resistance would otherwise be problems.

Radiation detection transducers

A **pyrometer** is a temperature-sensing device which responds to radiant energy from a target body. It therefore differs radically from all the other types of temperature sensor we have considered, which all rely on direct contact with the body or substance being measured. Pyrometers clearly have the advantage, therefore, where very high temperatures, which would damage other sensors, or where moving bodies are involved.

The pyroelectric element within the pyrometer is a thin ceramic slice, typically of a lead zirconate titanate compound. During manufacture the slice is heated to just below the Curie temperature in the presence of an electric field, so that the crystal dipoles within the material become aligned with the applied field, and, as it cools, the slice retains the induced polarization. Electrodes formed on the two faces acquire a charge, the magnitude of which is governed by the internal properties of the polarized material, and, as the temperature of the slice increases, the polarization varies so that the captive charge at the surface of the material decreases. This leads to an imbalance in the induced charges at the electrodes, and the voltage across the slice increases. Thus the element can be considered as a radiation-sensitive capacitor, though it is effectively shunted by a high, non-linear resistance. The signal it develops is proportional to the change in temperature, and for a small change, δT, the voltage change is given by

$$\delta V = \frac{\lambda A}{C} \delta T$$

where λ is the pyroelectric coefficient of the material, A is the cross-sectional area of the slice and C is the capacitance between the electrodes.

The voltage developed by the element decreases as the frequency of the radiant energy increases, and pyrometers are designed to operate over a specified range, normally within the infrared region, at wavelengths between 1 and 15 μm. The element encapsulation includes a quartz window to allow the radiation to reach the element, and the window properties are chosen to transmit either a broad range of frequencies or, with a 'daylight' filter, a narrower range which excludes the short-wavelength infrared contained in sunlight. Also included in the package in most cases is an n-channel FET which is used as a low-noise impedance-matching device between the pyroelectric element and the signal preamplifier, as shown in the circuit of Fig. 2.24.

Because they do not require contact with the radiating body, infrared

Fig. 2.24 Typical circuit for a pyrometer

temperature detectors are useful in a variety of applications. A single detector, for example, can be used to map the variation of temperature over a large surface. It could also be used to detect the passing of an object such as a billet in a steel mill, or to detect the presence of an intruder in protected premises. Infrared detectors are used in equipment to measure the concentration of certain gases which absorb infrared energy. The gas, such as carbon monoxide, carbon dioxide or methane, is led through the infrared beam of known intensity and the resulting signal at the detector indicates the density of the gas. This technique has been adapted to measuring visibility at airports.

Optical sensors

Reflective opto-switch

Slotted opto-switch

Miniature infrared sources and sensors are readily available as discrete components or housed together in suitable packages. The source is a highly efficient gallium arsenide light-emitting diode (LED) giving a maximum output of about 1 mW at 940 nm wavelength. The detector is an n-p-n silicon phototransistor or photo-Darlington connection. In a reflective opto-switch the source and detector are mounted side by side and use reflected energy to detect a body up to about 5 mm from the switch. In the slotted opto-switch the source illuminates the sensor directly and the slot allows an object, such as a rotating disc, to interrupt the beam. A filter is normally included in the packaging to reduce interference from extraneous light, and also to protect the device from dust and dirt. These switches are widely used in limit-switching, event-counting and optical-encoding applications.

An extension of this idea leads to the proximity detector, in which a built-in GaAs LED generates a modulated infrared light beam which is reflected from an external surface and detected by a sensitive sensor also built into the package. By adjusting the sensitivity, the device can be made to give a binary signal when suitable objects are at a defined distance (usually up to a few centimetres). The L-SER OS-1F from Koden Industry Co. Ltd and the ML 4-8-H-KSU from Visilux Ltd are typical of these devices. One of the most common uses for optical sensors nowadays is for the reading of bar codes. Hewlett-Packard, amongst others, manufacture optical reflective sensors specifically for this purpose, the HEDS-1000 being one such device, mounted in a TO-5 header.

In certain cases it is necessary to detect visible light, and we would then use silicon photodiodes doped to give good response in the visible range, or photovoltaic or photoconductive devices. The ORP12, for example, is a common photoconductive resistor using a cadmium sulphide cell which reduces in resistance as the incident light intensity increases. The dark resistance is very high, at about 10 MΩ, but at 1000 lux incident energy, that is 50 μW/mm, the resistance falls to only a little above 100 Ω. A typical use of such a device is shown in Fig. 2.25, in which the relay, R, operates when the light energy exceeds a certain level preset by the 5 kΩ resistor. The very much slower speed of reaction to light changes of the light-dependent resistor (about 100 ms) when compared with photosemiconductor devices (of the order of 1 μs) is of no consequence in this application.

The **linear array** (or line-scan) sensor is a monolithic image-sensing device where the photosensitive cells are based on the **charge-coupled device** (CCD) or photodiode technology, and are arranged as a single line of elements. For CCD sensors this can extend from 256 to 2048 active elements. A typical sensor, such as the Fairchild CCD111, consists of 256 sensing elements of cell size 13 μm by 17 μm with a spacing of 13 μm between cells. These sensors are commonly used in robotic and imaging applications for the detection of objects in two dimensions (two, because if the object is moving, or the sensor is swept along by, say, a robot end-effector, a 'shadow' moves across the sensor and the number of cells, or pixels, illuminated or not gives an indication of the area of the object). The line-scan array is mounted in a camera provided with a focusing lens to give a sharp image. The linear array can be considered electrically as a shift register. Low-resolution (30 × 30 pixels) cameras have been constructed based upon an MOS shift register, the Advanced Micro Devices Am2806. The Am2806 is a 1024-bit register fabricated in pMOS. The active bit elements are arranged as a 'meander' line

See J. J. Sparkes, *Semiconductor Devices*.

'Low resolution vision sensing', D. G. Whitehead and A. G. Mason, *Microprocessing and Microprogramming,* Vol. 16, No. 4–5, 1985.

Fig. 2.25 A light-operated relay

of 30 rows with the number of bits per row varying between 33 and 35. A usable square array of 30×30 pixels can be obtained from these.

Sonic transducers

Some of the applications mentioned above can make use of sound energy rather than electromagnetic energy. Sound transducers operate in the ultrasonic or inaudible range, which is commonly taken to include any frequency above about 18 kHz but in practical terms implies a frequency of about 40 kHz. The mechanical movement which causes the sound pressure waves is generated by either magnetostrictive action in a nickel alloy core, or piezoelectric action in a quartz crystal or lead zirconate titanate material. Magnetostriction is very similar to the piezoelectric effect except that it is the magnetic properties of the material which change under stress. Electrical energy is supplied to the transducer coil and the core transforms it to mechanical movement. Piezoelectric transducers are smaller and more efficient but magnetostrictive tranducers can handle much greater power, and are therefore preferred in specialized applications such as the hydrophone used in underwater sound processing. With the small, commonly available piezoelectric transducers the transmitter radiates continuous or modulated waves in a 20° cone, and the receiver can operate successfully over a few metres range, so these transducers are appropriate for local remote-control or data-transmission systems. At very much higher frequencies, that is above 10 MHz, the wavelength becomes very small relative to everyday objects, and the sound beams can be focused and manipulated in the same way as light beams. These very high frequency sound waves are widely used in medical and dental equipment.

Nuclear radiation detectors

Special radiation detectors are required to measure the rate at which nuclear radiation is received. The radiation can consist of alpha particles – helium nuclei – or beta particles – high-speed electrons or positrons – and the sensing principle is to detect the ionization of a gas in an ion chamber caused by an arriving particle colliding with a gas molecule. The best-known instrument is the Geiger counter which consists of two electrodes in an envelope containing the gas. The electrodes have a sufficiently high potential difference maintained so that any ion created by an incoming particle is accelerated and produces other ions by collision. The resulting charge surge at the electrodes is used to generate a voltage in the external circuit. Gamma rays – shortwave X-rays – can also cause ionization in the chamber.

Chemical activity

In order to measure chemical activity we must find some link between the strength of the activity and an electrical signal, and in many cases we can do that by using the concentration of ions existing in a solution. For example, the control of the acidity of a solution is very important in many industrial and horticultural operations, and the degree of acidity is determined by the concentration of positively charged hydrogen ions, H^+, present in the solution. This is known as the hydrogen potential, pH, value. A molar solution of a strong acid, such as hydrochloric acid, contains one gram of hydrogen ions in each litre of solution; a molar solution of a strong alkali has 10^{-14} grams of H^+ ions per litre; pure water is a neutral solution which has a strength of 10^{-7} grams per litre. The pH value is defined as

A molar solution of a substance contains its molecular weight in grams per litre of solution.

$$pH = -\log[H^+]$$

where $[H^+]$ is the hydrogen ion concentration in grams per litre. Thus the pH value ranges from 0 for a strong acid, through 7 for a neutral solution, to 14 for a strong alkaline solution.

A sensor for pH value measurement is constructed in the form of a probe which has a porous glass membrane at its tip. Hydrogen ions in solution diffuse through the membrane and react with lithium ions, which are contained in the membrane, to generate a potential which is proportional to the hydrogen ion concentration. A reference potential is maintained by an internal element consisting of a silver–silver chloride wire surrounded by a gel of known pH value. The probe output is in the form of an emf developed across a very high impedance, typically over $100\,M\Omega$, so special high input-impedance amplifiers must be used. The sensitivity of a commercial probe is of the order of 60 mV per unit pH.

A different sort of reaction is utilized in the platinum wire gas sensor used in detecting the presence of certain gases, including natural gas, methane and propane. These sensors make use of the **Pellistor principle**, in which the resistance of certain chemicals changes when gas molecules are absorbed at the surface. The platinum wire is given a coating of such a material, with an amount of catalyst included, and this is heated to its operating temperature by a current flowing through the wire. In order to compensate for changes occurring in the wire due to fluctuations in ambient temperature and humidity levels, a second wire is included in the sensor which is identical to the first except that it is not coated with the sensitive material. The two elements are used as components of a bridge (Fig. 2.26), which is initially balanced by

Instrumentation amplifiers are dealt with in Chapter 3.

Fig. 2.26 Bridge circuit for a gas sensor

adjusting the potentiometer. The elements are mounted side by side on a suitable header, and are covered with a fine-mesh wire netting for protection, but also to prevent explosions. The sensor output voltage remains essentially linear for gas concentrations up to 10 000 ppm but is dependent on the gas, the output for propane, 20 μV/ppm, being more than twice that for methane.

10 000 ppm is equivalent to 1%.

Actuators, stepper motors and displays

Output transducers are divided into two main categories for convenience, though as always it is difficult to draw a distinct line between them. We will define an actuator as a device which responds to an electrical signal and develops an output in another form, usually mechanical movement of some sort. They are used mainly in controlling flow or position. Displays, on the other hand, convert the electrical signal into a form which provides information to an 'observer'. The vast majority of displays are therefore optical, in that the input signal is converted to a light output. However, the optical output is, in some cases, provided by the positioning or rotation of vanes, for example, which means that the display itself includes an actuator. Similarly, the 'display' might be a loudspeaker, which converts the electrical signal to sound energy, but when used in a vibration tester the same loudspeaker action must come within our definition of an actuator.

Actuators

The conversion of electrical signals into mechanical movement, i.e. electro-magnetomechanical energy conversion, usually employs an unmagnetized iron member which moves in the direction of a magnetic field generated by the applied electric signal. The solenoid introduced in Chapter 1 is a simple example which exists in many shapes and forms. One form is the relay, a control device whereby the electrical signal can be used to make or break one or more separate electrical circuits. Figure 2.27 shows the essential elements of solenoid construction and how the moving iron action is utilized to operate switch contacts in the relay.

The work done in moving the plunger a distance dx is the product of force and distance:

$$W = f \, \mathrm{d}x$$

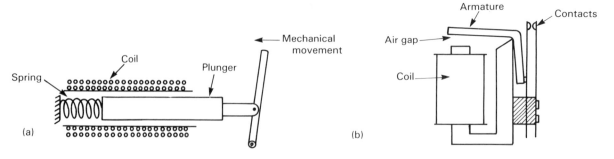

Fig. 2.27 (a) The solenoid. (b) The relay

The movement of the plunger in the coil causes a change in inductance of dL, giving a change in energy of $\frac{1}{2}I^2\,dL$ joules. Thus

$$f\,dx = \tfrac{1}{2}I^2\,dL$$

or

$$f = \tfrac{1}{2}I^2\,dL/dx \text{ newtons}$$

where I is the applied current in amps, dL is the change in inductance in henries and dx is the distance moved in metres.

Work is W joules, force is f newtons (N). The energy stored in an inductance is $\frac{1}{2}LI^2$.

An excellent treatment of this appears in *Circuits, Devices and Systems*.

Worked example 2.6

The variation of inductance with depth of penetration of a plunger in a solenoid is shown in Fig. 2.28. Estimate the range over which the maximum pull can be achieved, and its value, if the current through the coil is 1 A.

Since the force exerted on the plunger is proportional to the rate of change of inductance with distance, the maximum pull is achieved when dL/dx is a maximum. From the graph, the maximum slope, and therefore the maximum pull, exists in the range 1 to 3 cm, and has a value

$$\frac{dL}{dx}_{(max)} = \frac{0.3 - 0.1}{(3 - 1)10^{-2}} = 10 \text{ H/m}$$

With a current of 1 A, the force is given by

$$f = \tfrac{1}{2} \times 1^2 \times 10 = 5 \text{ N}$$

As already noted, the distance moved by the plunger is very small, as the force exerted by the coil drops off rapidly with distance, and for longer mechanical movement a lever action can be employed. The relay in Fig. 2.27 has a free soft-iron armature which, when attracted towards the solenoid pole-piece, operates a lever which, in turn, closes, or opens, electrical contacts. By use of such relays, heavy currents or high voltages can be controlled by relatively low-level signals. The force on the relay armature is given approximately by

Fig. 2.28 Variation of inductance with solenoid plunger displacement

$$f = \frac{\mu_0 N^2 I^2 A}{2l^2} \text{ newtons}$$

where N is the number of turns on the coil, I is the current in amperes, A is the area of the airgap in square metres, l is the length of the airgap in metres and μ_0 is the permeability of free space ($4\pi \times 10^{-7}$ H/m).

The reed relay, or magnetic reed switch, consists of two slivers, reeds, of a ferromagnetic material, such as nickel-iron, hermetically sealed into a glass tube, with the ends of the reeds aligned and a small gap between them. The whole assembly is inserted into a coil, and, under the influence of the magnetic field set up when current flows in the coil, the reed ends are mutually attracted. When the field collapses, the reeds spring apart. The inertia of the reeds is low and fast operation is possible, some reed switches operating in less than one millisecond. The operating speed is usually limited, in fact, by contact bounce (illustrated in Fig. 2.29). This can become progressively worse during the life of the reed, resulting eventually in unreliable switching and possibly complete failure. The contact ends of the reeds are plated with a precious metal such as rhodium to provide low contact-resistance, and to extend their life, which is measured in millions of operations. To achieve the fastest switching times and most reliable operation, contact bounce is eliminated by the use of mercury-wetted contacts; the surface tension of the mercury film between the contacts maintains a bridge of mercury and ensures immediate, unbroken contact.

This is true of any switch.

Reed switches are also used in conjunction with permanent magnets. Burglar alarm relays on window frames, for example, commonly use a permanent magnet set into the moving part of the window and a reed switch embedded in the frame. When the window is shut, the reed switch is held closed, but opening the window moves the magnet away and the reeds spring apart.

Fig. 2.29 Contact bounce and its control by use of two cross-coupled NAND gates

Typical ratings: coil voltage 3.7–10 V dc
contacts 240 V ac
200 V dc
10 W
Coil resistance 500 Ω

Fig. 2.30 Outline of reed switch dual-in-line package

48

Because of their small size, little power is required to operate reed switches and the coil can often be driven directly from transistor–transistor logic (TTL) circuitry. In fact, some reed switches, complete with coil, are built into standard dual-in-line packages for ease of handling with logic devices in similar packages (Fig. 2.30).

In use, solenoids must be treated as inductors, and for correct operation consideration must be given to protecting against the self-induced, or back emf generated when the primary excitation circuit is broken. When the transistor in the circuit of Fig. 2.31 conducts (when the switch is closed), the current builds up through the solenoid according to the relationship

$$i = V/R_{\mathrm{s}}[1 - \exp(-R_{\mathrm{s}}t/L)]$$

The ratio L/R_{s} is the time constant of the circuit. If L is in henries, R in Ω, then L/R_{S} is in seconds.

The greater the value of the series resistance, R_{s}, relative to the inductance, L, the faster will the current reach its maximum value, V/R_{s}. In applications demanding a very fast response, such as the needle solenoids in impact printers, the solenoids are often driven from a high voltage source via a relatively high series resistance. Considerable power is dissipated in the resistor but the solenoid response time is improved enormously. When the transistor is turned off (the switch is opened), the current flow is arrested and the magnetic field collapses, inducing an emf across the inductance in a sense that seeks to maintain the current. If the cessation of current flow could take place in zero time then the induced emf would, in theory, have an infinitely large value for an infinitely short time. In practice, the current cannot stop abruptly but large voltages can be created. If the opening switch is in the form of a mechanical contact, as is the case in the ignition circuit of a car engine, then a spark occurs across the contacts and a decaying current flows for a short period. The contacts must be designed to withstand the surface erosion this arcing causes. Similarly, if the switch is a transistor, protection must be given to prevent the back emf from destroying the device. The usual method is to connect a clamping diode across the coil (Fig. 2.32), to provide a low-resistance path to the induced emf, and prevent the voltage rising above the supply voltage level by more than the forward volt drop of the diode. All

$e = -L\,\mathrm{d}i/\mathrm{d}t$. A current of 1 A through an inductance of 1 H, terminated in 1 μs, gives an emf of 10^6 V!

Fig. 2.31 A solenoid controlled by a transistor

Fig. 2.32 Diode protection for the solenoid or relay switch

engineering involves compromise, and in this case the penalty paid for safety is an increase in recovery time of the solenoid. The collapsing magnetic field causes current to flow through this low-resistance path with a magnitude

$$i = I_0 \exp[-(R_s + R_f)t/L]$$

where I_0 is the initial current (V/R_s if we ignore any voltage drop across the switch), R_s is the resistance of the coil and R_f is the forward resistance of the diode. The net effect is that the field is maintained longer, causing the armature, or plunger to be released slowly. This recovery time is dependent on the various circuit parameters.

Worked example 2.7

In the circuit of Fig. 2.32, the supply voltage is +12 V, the coil resistance is 10 Ω and the coil inductance is 0.5 H. The forward resistance of the diode is 8 Ω. If the closure current of the relay is 0.5 A, and the release current is 0.1 A, what are the make and break times?

The time taken for the relay to operate when the transistor is turned on is found from the equation

$$i = V/R_s[1 - \exp(-R_s t/L)]$$

that is,

$$0.5 = \tfrac{12}{10}[1 - \exp(-10t/0.5)]$$

whence

$$\exp(-20t) = 0.584$$

and

$$t = 26.9 \text{ ms}$$

When the transistor is switched off, the time for the current to fall to 0.1 A
is found from the equation

$$i = I_0 \exp[-(R_s + R_f)t/L]$$

which gives

$$0.1 = \tfrac{12}{10} \exp[-(10 + 8)t/0.5]$$

so

$$\exp(-36t) = 0.083$$

and

$$t = 69 \text{ ms}$$

It is possible to improve the response time of a solenoid by using a parallel
damping resistor rather than a diode. In this case, the value of the resistor
must be chosen to prevent damage from the back emf and to provide the
required recovery time. These often conflicting demands can be met by the
use of a non-linear resistor, such as the voltage-dependent resistor (VDR).
This device exhibits a high resistance until breakdown potential is reached,
whereupon the resistance falls to a very low value. The breakdown potential
lies between 50 and 400 V, depending on the component used.

Applications involving the driving of solenoids or relays from a dc supply
can conveniently make use of the drivers packaged in standard dual-in-line
format. The Sprague ULN2803A, for example, contains seven independent
drivers, each capable of switching up to 0.5 A and controlled by TTL level
signals. Clamping diodes are provided with each driver. If the power source is
ac, the triac can be used as the control element. For reliable operation in this
case it is desirable that the switching on and off of the solenoid be performed
at the zero-crossing point, that is when the power source sinusoid passes
through 0 V. The switching transients can then be avoided. The circuit of
Fig. 2.33 illustrates the method and uses an edge-triggered D-type flipflop.

See the delay flipflop, Chapter 4
of *Digital Logic Techniques*.

Fig. 2.33 Zero crossing point control of an ac power source using an
edge-triggered D-type flipflop (e.g. 4013B)

The **stepper motor** is a very versatile device widely used in converting digital information into proportional rotational movement. Unlike the dc motor, the stepper motor has a rotor, constructed from permanent magnets, which rotates in discrete steps around a multi-pole stator, following a rotating field generated in the stator. The rotor is, in effect, pulled from stator pole to stator pole and so is capable of controlled motion, both in the number of steps and in the stepping rate. These motors are used extensively in digital control systems, and uses range from positioning of robot arms to printer mechanisms. Through appropriate screw gearing the rotational action can be translated into accurate linear movement for X–Y plotters and similar equipment.

The principle of operation can be explained with reference to Fig. 2.34. The rotor is a permanent magnet, constrained to rest in the position shown in Fig. 2.34a by the magnetic field pattern produced by the stator pole-pieces. If I_1 is reversed, the rotor is forced into a new position, Fig. 2.34b. By switching the polarity of the stator currents in the sequence $+I_1$ $+I_2$, $-I_1$ $+I_2$, $-I_1$ $-I_2$, $+I_1$ $-I_2$, so producing a rotating field pattern, the rotor is moved through 360°. The speed of rotation of the rotor is governed by the rate at which the stator currents are switched, and the direction of rotation by the sequence used. Maintaining the currents constant at any point in the sequence holds the rotor stationary. The arrangement shown is a 2-pole (rotor), 4-phase (stator) motor but, in practice, most stepper motors have two sets of stator coils with many pole pairs arranged around the rotor. Two-coil motors require bipolar push–pull drive circuits but a more common arrangement, called **unipolar** drive, has two coils wound on each stator bobbin. On reversing the connections to each coil the stator flux is reversed, as one or other of the coils is energized from the single power supply. Such motors are said to be **bifilar wound**. The angle through which the rotor moves each step is known as the **step angle** and this is always a submultiple of 360°. Our simple example in Fig. 2.34 has a step angle of 90° but more typical values are 7°30′ and 3°45′. An important feature of a stepper motor is that any

(a)   (b)

Fig. 2.34 The stepper motor

positioning error is dependent only upon the accuracy of pole placement within the motor and is non-cumulative. Positioning errors average out to zero within the four-step sequence needed to move the rotor one pole pitch. Small stepper motors can develop torques of up to a few newton-metres, but the torque falls off with increasing stepping speed and a maximum of a few hundred steps per second is to be expected. The maximum speed of the motor is limited by hysteresis and eddy current losses. If the motor is too heavily loaded it can lose steps and, hence, its accuracy. To avoid this, and to achieve maximum possible acceleration under particular load conditions, stepper motors are often 'ramped' up to speed (and down again). The stepping rate is made variable and is controlled so as to increase to the desired rate as the motor speed increases. As the rate at which the current can be developed in the stator coils is limited by the inductance of the windings, for maximum performance a constant-current drive is preferable, and switching the currents through resistors from a relatively high voltage source is the most effective way of achieving this.

The generation of the control sequence for the stator currents is achieved most economically for smaller motors by the use of one of the standard integrated circuits and by use of commercial drive cards for the larger, more powerful motors. The Signetics SAA1027, for example, is an integrated circuit designed to drive a 4-phase stepper motor directly up to a maximum of 350 mA per phase. It interfaces with TTL circuitry and generates all the necessary stator current sequences from an external clocking pulse train, requiring only one other signal to indicate by its level the direction of rotation. SGS-Thompson supplies integrated circuits capable of driving both unipolar and bipolar motors. For the larger stepper motors manufacturers normally provide suitable driver systems, and these systems include ramping circuitry so that the motors may be accelerated quickly and accurately.

Visual displays

Visual displays are used to indicate either qualitative or quantitative events. By a qualitative event we mean the presence or absence of some on/off signal, and such indications are usually provided by an incandescent lamp or light-emitting diode (LED). LEDs are the more popular because of their superior reliability. The LED consists of a semiconductor p−n junction fabricated from materials such as gallium arsenide and gallium phosphide. Under the correct bias conditions, injected minority carriers recombine in the junction region with the emission of light energy at a wavelength corresponding to the width of the forbidden energy gap. Red, yellow and green LEDs are available, with red being the most common.

Infrared LEDs are also available and are widely used in detection circuits as the source to a spectrally matched silicon photodiode receiver.

In use, the LED is typically operated simply as a forward-biased junction diode, as in Fig. 2.35. Since adequate light output of 2 to 4 mcd can be obtained with around 8 mA of forward current, LEDs can be driven directly from low-power TTL operating as a current sink. More current is necessary to give brighter displays, and a buffer gate, such as the 7406 should be used. It is interesting to note that driving the LED with large current pulses, illustrated in Fig. 2.35c, gives a much increased light output but a smaller average power dissipation in the diode.

The 'candela' is the luminous intensity, in a given direction, of a monochromatic source at 540×10^{12} Hz, with a radiant intensity of $1/683$ watts/steradian.

Fig. 2.35 (a) LED as a power supply indicator. (b) LED controlled from a TTL gate. (c) Using a higher level pulsed current

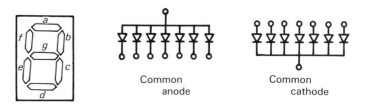

Fig. 2.36 The seven-segment display

Alphanumeric displays provide both numerals and letters, and the simplest form is the seven-segment display (Fig. 2.36). Each segment consists of a translucent plastic bar containing a single LED, and, by lighting the appropriate bars, any numeral and a few letters can be displayed. More complex bar patterns are provided when additional letters are to be displayed, or LEDs forming a dot matrix (often 5×7) are used. Bar graph displays are also available. Electrically, the diodes have either all of their anodes or all of their cathodes connected together. It is not necessary to design drive circuits for these displays as specialized circuits are readily available. The 74LS47 TTL decoder, for example, accepts binary coded decimal (BCD) input and drives a seven-segment display directly, with a capability of sinking up to 24 mA per segment. The circuit is given in Fig. 2.37.

Binary coded decimal is explained in *Computers and Microprocessors*.

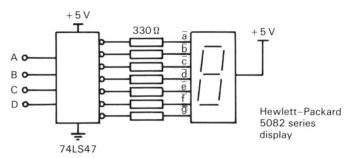

Fig. 2.37 Driving a seven-segment display

54

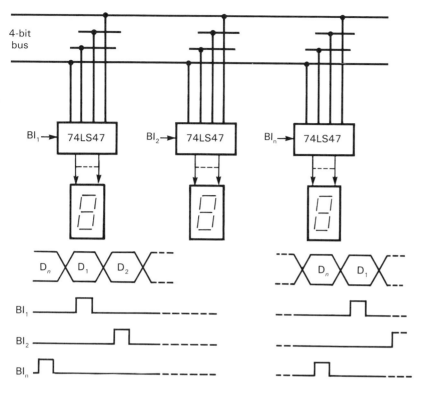

Fig. 2.38 A multiplexed display

When several digits are to be displayed, multiplexing can be used to reduce circuit complexity and to take advantage of the pulse mode characteristics of the LEDs. A multiplexing circuit is shown in Fig. 2.38, making use of the 74LS47 input labelled BI. This is a blanking input and when taken low it switches off the display. Thus, if we arrange for the BCD codes to be generated sequentially, and at the correct time we unblank the appropriate display, the persistence of vision of the observer's eye gives the impression of a steady multiple-digit display. The repitition rate of the pulse sequence and the level of current determine the effective brightness.

The Hewlett-Packard HDSP-5500 series seven-segment displays have a quoted output intensity of 2.5 mcd for an input current of 10 mA dc. If the 74LS47 driver can supply a maximum pulsed output current of 50 mA (for a duty cycle less than 50%), determine how many displays can be driven at this current level if a refresh rate of 1 kHz is to be used. What is the peak intensity of the display?

From the graph, a peak current of 50 mA is allowable if the duty cycle is kept down to 12%, i.e. a pulse width of 0.12 ms. Therefore a total of $1/0.12$, i.e. eight displays can be pulsed within the 1 ms refresh period. For 50 mA pulses, the luminous intensity will be $2.5 \times 1.3 = 3.25$ mcd.

Worked example 2.8

Characteristics for a typical 7-segment display showing maximum pulse width and relative efficiency as related to peak current. The efficiency is shown relative to a dc current of 10 mA and the pulse width to a refresh rate of 1 kHz.

See Chapter 6.

The generation of the blanking signals can also be multiplexed, thereby reducing the number of signal lines further. Because of the continuous demands of the display, this mode of operation, though simple, is not ideal for microprocessor-based systems since many other jobs must be dealt with at the same time. It is more appropriate for a dedicated hardware arrangement where the data would be derived from a series of multiplexed latches. Displays such as the Hewlett-Packard HDSP-2000 series are supplied with integral latches, simplifying the design still further.

An alternative method of driving multiple displays, particularly useful in microprocessor applications, uses a special integrated circuit to accept a serial input data stream and provide the correct current drive to each display segment directly. The MM5450N, for example, has 34 outputs: serial data, formatted as 34 data bits preceded by a 'start' bit at '1' and synchronized to a clock signal, is shifted into the chip. When all bits have been shifted in, the data bits are latched and held until the next 35 bits are received. The display current is controlled by a single external resistor, defining a current of approximately 1/20 of one output segment current. A simple switching circuit on the current input can allow a pulsed, higher current level to be used, giving increased brightness without increasing the average dissipation.

Where low power consumption is essential, liquid crystal displays are preferred since they draw very little current; typically less than 10 mA. Basically, the LCD relies on the ability of the very long molecular chains existing in the crystal to align themselves in the presence of an electric field. Under the correct conditions, the unaligned crystal reflects light, but when energized, the aligned crystal becomes transparent to light. The electric field is applied across a sandwich of the liquid crystal held between parallel plates of glass coated with a transparent conducting layer. Because these displays reflect or transmit light, rather than generate light, unlike LEDs they can be used successfully in direct sunlight. LCDs that transmit light are usually supplied with an electroluminescent backing so that they can be viewed without an external source. The more complex drive requirements of the liquid crystal displays are met by special-purpose devices which provide not only the decoding circuitry but also the necessary alternating voltage generators. The Intersil ICM7211, for example, is a four-digit seven-segment LCD driver. Data input can either be multiplexed BCD (ICM7211), using separate digit select inputs, or be provided with chip select controls and address latches (ICM7211M), for microprocessor applications. Note that there is a functionally equivalent LED driver, the ICM7212/ICM7212M.

Because of their cost, LCDs and LEDs are usually more appropriate for small displays, but large arrays of LEDs have been used to good effect in public display boards such as the noticeboards on the London Underground. The larger displays used in public announcement panels at railway stations, airports and sports stadiums, often make use of a magnetic device similar to that shown in Fig. 2.39. The permanent magnet in the rotor is attracted or repelled by the solenoid magnet, depending on the direction of the current, thus rotating the vane. The rotor is stable in either position, and current is needed only to change from one to the other. The vanes are of lightweight material, coated in fluorescent paint against a black background. These devices have been built into seven-segment format displays over 600 mm high,

Fig. 2.39 Seven-segment devices for large displays

and good visibility at up to 300 m can readily be obtained.

Summary

The number of transducer types is almost unlimited, and in order to bring our area of study down to a more manageable size we have considered transducers under four main headings. Input transducers for detecting mechanical change allow us to sense force, pressure, position, proximity, displacement, velocity, acceleration, vibration and shock in all their multiple manifestations. The basis of many mechanical sensors is the strain gauge which is usually used in a bridge configuration. Other devices such as the LVDT and synchro are also widely used. Temperature transducers form another large group, and we have looked at the operating principles of the major types, with some of the techniques used in compensating for non-ideal characteristics. Radiation and chemical sensing transducers form the remaining groups. Actuators rely almost entirely on electromagnetic action and, in modern equipment, occur most commonly as solenoids and relays, including the reed relay, and stepper motors. Visual displays also come in a bewildering range of types and sizes, but, because of their ease of interfacing with electronic circuitry, the majority are based on the LED and LCD.

Review questions

1. What is meant by gauge factor?
2. Define Young's modulus.
3. What is meant by the time constant of a solenoid? A 24 V solenoid of 2 H inductance takes a current of 1 A. What is its time constant?
4. Describe the operation of a thermocouple. Under what conditions would a radiation pyrometer be more appropriate for the measurement of temperature?
5. Describe the operation of the LVDT.
6. What are the synchro format voltages?

Further reading

1. *Stepping Motors; Guide to Modern Theory and Practice* (2nd edn), P. P. Acarnley, Peter Peregrinus Ltd, 1984.
2. *Measurement Systems*, E. O. Doebelin, McGraw-Hill, 1983.
3. *Computers and Microprocessors: Components and Systems,* A. C. Downton, Van Nostrand, 1984.
4. *Circuits, Devices and Systems*, R. J. Smith, Wiley, 1984.
5. *Digital Logic Techniques – Principles and Practice* (2nd edn), T. J. Stonham, Van Nostrand, 1987.
6. *Transducers in Digital Systems* G. A. Woolvet, Peter Peregrinus Ltd, 1979.

Problems

2.1 A 2 H solenoid is connected in series with a resistor so that the total resistance is 12 Ω. Sketch the graph of current against time when a 12 V dc supply is applied.

2.2 The plunger pull-in current for the solenoid in question 2.1 is 0.5 A, and its drop-out current is 0.3 A. If the solenoid supply voltage is switched between 0 V and +12 V with a mark-to-space ratio of 1 : 1, i.e. equal on and off times, and a total period of 1 s, what is the plunger operating mark-to-space ratio?

2.3 What is the gauge factor of a 200 Ω conductor that is 25 mm long if, under a tensile force, the resistance changes by 12 Ω and the length changes by 0.5 mm?

2.4 A force transducer uses a Wheatsone bridge circuit consisting of four 250 Ω strain gauges arranged as a two-active, two-passive network. The excitation voltage is +10 V dc. When a force of 2 N is applied, an output voltage of 40 mV is developed. What is the change in gauge resistance?

2.5 An LVDT has the following characteristics:

+ core displacement (mm)   0.05 0.10 0.15 0.20 0.25 0.30 0.35 0.40 0.45 0.50 0.55

+ output voltage (V)   0.065 0.135 0.205 0.270 0.340 0.410 0.475 0.535 0.585 0.625 0.660

The excitation voltage is 5 V at a frequency of 1 kHz. Calculate the sensitivity in mV mm^{-1} V^{-1} excitation. The quoted linearity is ± 10 μm; what is the range of core displacements?

3
Analogue processing of signals

Objectives

☐ To indicate the need for processing of analogue signals.
☐ To summarize the performance of ideal and practical op-amps.
☐ To consider the performance of other specialized amplifiers.
☐ To introduce problems caused by electrical noise and methods of minimizing their effects.
☐ To explain methods of amplitude and frequency modulation of carrier systems.
☐ To describe the operation of the analogue scanner.

Introduction

In this chapter we restrict ourselves to the most common types of sensor systems; that is, those in which the output signal is electrical in nature. Such an output signal may be a voltage or current delivered either directly from the sensor or from a bridge circuit, of which the sensor forms a part. We have seen that sensors are based on a very wide range of physical properties and it is to be expected that the electrical signals which are generated are equally widely varying in magnitude and range. In addition the effective output characteristics vary considerably, dependent on the type of sensor. In most cases it is necessary to amplify the raw signal before it is used or further processed, and even when it is not strictly necessary, it is often worthwhile using an amplifier to allow adjustment of the impendance levels which are seen by the subsequent circuitry. For example, a commonly used capacitor-based microphone, the electret, has an FET-follower circuit mounted in the microphone itself. Although this amplifier has a voltage gain less than 1 it has a very high input impedance and a low output impedance. The result is that the capacitive element, with its intrinsically high output impedance, is usable with long lengths of cable when the cable capacitance would otherwise have resulted in a severely reduced output and restricted frequency range.

The effect of cable capacitance can also be important when a cathode ray oscilloscope (CRO) is used, since the capacitance of the interconnecting leads from the CRO to the point of measurement can reduce its effective impedance to a very low value, thus loading, and possibly distorting, the measured signal. This loading effect is more apparent where high-frequency components are present, especially where the high-speed edges of digital

For gm = 2mS
$R_o = \frac{1}{gm} = 500\,\Omega$

See Ritchie, G.J. *Transistor Circuit Techniques* (Van Nostrand Reinhold, 1983), p. 143.

Hand-held microphone

59

voltage waveforms are to be measured. To compensate for the additional shunt capacitance of the cable we often use a passive compensated probe at the circuit end of the cable. This consists of an adjustable coaxial capacitor in parallel with a fixed resistance, which inevitably leads to an attenuation of the signal, so the resistance value is chosen to give a convenient attenuation factor. A factor '× 10' is usually used so that the effect is the same as switching the oscilloscope input attenuator to its next higher position. If the additional attenuation is unacceptable, an active probe must be used in which an amplifier, similar to the microphone circuit, is mounted within the probe, presenting a high impedance to the measurement point and a low driving impedance to the capacitive connecting cable.

Now, the equivalent circuit of the complete arrangement must take account of the probe components, R_p and C_p, the effective imput resistance and capacitance of the oscilloscope, R_i and C_i (typically 1 MΩ and 20 pF), the capacitance of the probe cable, C_c (typically about 25 pF/m), and other stray capacitance, C_s (typically a few picofarads). From the potential divider action of $C_p \| R_p$ and $(C_c + C_i) \| R_i$, we get

‖ indicates 'in parallel with'.

$$\frac{V_2}{V_1} = \frac{R_i/(1 + j\omega(C_c + C_i)R_i)}{R_i/(1 + j\omega(C_c + C_i)R_i) + R_p/(1 + j\omega C_p R_p)}$$

If C_p is adjusted to give

$$(C_c + C_i)R_i = C_p R_p$$

the ratio becomes

$$\frac{V_2}{V_1} = \frac{R_i/(1 + j\omega C_p R_p)}{R_i/(1 + j\omega C_c R_p) + R_p/(1 + j\omega C_p R_p)} = \frac{R_i}{R_i + R_p}$$

This indicates that the attenuation is independent of frequency and under these conditions the probe is correctly compensating for the various capacitances, so that the signal is unaffected except in magnitude. For convenience, if R_i is 1 MΩ, R_p is chosen to be 9 MΩ and the gain of the network is then

$$\frac{V_2}{V_1} = \frac{1}{1 + 9} = \frac{1}{10}$$

This leads to the common name of a 'times ten probe'. The effective input capacitance of the network is

$$C = C_s + \frac{C_p(C_c + C_i)}{C_p + C_c + C_i}$$

which is around 10 pF for a typical modern probe.

The ideal operational amplifier

The most widely used type of amplifier for signal conditioning is the instrumentation amplifier, which is a composite circuit using several amplifiers to provide an accurately controlled amplification of the difference between the two input voltages. In order to understand the operation of the instrumentation amplifier, therefore, we must first review the action of a basic operational amplifier, assuming at this stage that it can exist in an ideal form. We then consider the other specialized amplifiers and processing methods.

The operational amplifier, or op-amp, is so named because it was developed initially to perform the mathematical operations of addition and integration in analogue computers. It is now probably the most widely used analogue electronic circuit, in most cases fabricated as a single silicon integrated circuit. Some very highly specialized circuits are constructed in hybrid form; that is, to achieve the desired performance, both integrated circuits and discrete components are used together, usually mounted on a small ceramic substrate. The op-amp is, in fact, a difference amplifier with a very high intrinsic gain and high input impedance. Being a difference amplifier, its output signal is an amplifier form of the difference between the voltages existing at its two input terminals. The characteristics of the op-amp are chosen so that it can most readily be used in a feedback arrangement, with the overall characteristics being determined by a few components external to the op-amp. It is to achieve this that the gain without feedback (that is, the *forward gain* or *open-loop gain*) must be very high (ideally infinite), the input resistance very high (ideally infinite), and the output resistance very low (ideally zero). Commercial op-amps do approach this ideal, and allow us to formulate two rules of thumb which are very useful in analysing the action of a particular circuit. The maximum output voltage of the amplifier is determined by the dc supply voltages, and is, therefore, finite, so, as the amplifier gain approaches infinity, the difference voltage at the input must approximate to zero. Thus

Even the cheapest op-amps have a gain greater than 100 dB and an input resistance of several megohms.

> Rule 1: In a feedback amplifier, the differential input voltage, V_{dm}, equals zero.

A detailed presentation of the theory relating to this section is given in *Feedback Circuits and Op-amps*, Chapter 4.

The current drawn at the amplifier inputs is governed by the input resistance, and as the resistance approaches infinity the input current approximates to zero. Thus

> Rule 2: The op-amp input current equals zero.

The input is said to be at virtual earth or virtual ground.

If we now consider an op-amp circuit with negative feedback (Fig. 3.1a), we can use our two rules to determine its action. The feedback is negative since it opposes the input voltage. From rule 1, the inverting input is at virtual earth and $V_{dm} = 0$. The overall input resistance is R_1.

The current $i_1 = V_i/R_1$, and $i_2 = -V_o/R_2$. From rule 2, i_1 must equal i_2, so

$$V_i/R_1 = -V_o/R_2$$

(a)

(b)

Fig. 3.1 Op-amps with negative feedback: (a) inverting; (b) non-inverting which, when rearranged, gives

$$V_o/V_i = -R_2/R_1 = A_v$$

where A_v is the overall gain and the negative sign indicates a phase reversal in the amplifier. We note that if V_i increases, increasing the current i, the output voltage falls by an amount sufficient to increase the current i_2, so that i_1 still equals i_2. Similarly, if V_i decreases, the output voltage rises. Thus we can formulate a third rule of thumb applicable to the amplifier with feedback:

> Rule 3: The output voltage change resulting from an input voltage change is such as to maintain the difference voltage at the op-amp input at zero.

Hence the term 'virtual earth' in this configuration.

Figure 3.1(b) shows another form of the circuit in which the inverting input is earthed and the input voltage is applied to the non-inverting input. By potential divider action of R_1 and R_2,

$$V_x = V_o R_1/(R_1 + R_2)$$

and from rule 1,

$$V_x = V_i$$

so the overall voltage gain is

$$A_v = V_o/V_i = 1 + R_2/R_1.$$

This equation shows that the arrangement does not give the phase inversion of the previous circuit, and the application of rule 3 confirms it; if V_i increases, the output voltage, V_o, must also increase by an amount large enough to raise the voltage of point X by the same amount. This form of circuit further increases the very high resistance of the op-amp, and note, also, that both forms of the circuit in Fig. 3.1 dramatically reduce the already low output resistance of the op-amp. The effective input and output resistances of the circuit when feedback is applied are governed partly by the resistance values used and partly by the forward gain of the op-amp itself. In addition, however, the method by which the feedback signal is derived at the output, and the way in which it is introduced at the input, affect the output resistance and input resistance respectively. The feedback signal can be proportional to either the output voltage or the output current. Both the amplifiers of Fig. 3.1 derive a feedback signal proportional to the output voltage, and the effective output resistance of the circuit is then given approximately by

$$R_o = r_o/(1 + A\beta)$$

where R_o is the output resistance of the op-amp itself, A is the forward gain and β is the feedback fraction.

The feedback fraction, β, is determined by the feedback resistors and, for the inverting amplifier, $\beta = R_1/R_2$. The non-inverting amplifier gives a similar value of $\beta = R_1/(R_1 + R_2)$. In both cases, therefore, a typical circuit has an output resistance of a fraction of an ohm. The circuit of the inverting amplifier in Fig. 3.1a shows that the resistance R_1 is connected to the virtual earth at the input of the op-amp, and the effective input resistance is, therefore, simply $R_i = R_1$. With the non-inverting amplifier, however, the resistance seen at the input is increased by the effect of the output signal on the resistance R_1 in series with the inverting input, and the effective input resistance is increased to $R_i = R_1 (1 + A\beta)$. Here, R_1 is the differential input resistance between the op-amp input terminals.

Worked Example 3.1

The 741 is a widely used commercial op-amp, which has a typical forward gain of 160 dB. The differential input resistance, as measured between the two input terminals, is 2 MΩ, and the output resistance is 75 Ω. When negative feedback is applied, what values of R_1 and R_2 are necessary to give an overall gain of 40 dB, with phase inversion and an input resistance of 1 kΩ? What values of gain and input resistance would these same values of R_1 and R_2 give in a non-inverting form of the amplifier? What is the approximate output resistance in each case?

The inverting form of the amplifier, Fig. 3.1(a), gives an input resistance R_1. This is to be 1 kΩ, therefore R_1 equals 1 kΩ. The voltage gain, $|A_v|$, is R_2/R_1 and the value required is 40 dB. This is equal to a voltage gain of 100, so the value of R_2 necessary to give the required gain is 100 kΩ. Using these same values of R_1 (1 kΩ) and R_2 (100 kΩ) in a non-inverting configuration gives a voltage gain,

Gain in dB = $20 \lg (A_v)$

$$A_v = 1 + R_2/R_1$$

$$= 101$$

and an input resistance of

$$R_i = 2 \times 10^6 \left(1 + \frac{2 \times 10^5 \times 1000}{1000 + 100\,000}\right) \Omega$$

The forward gain, A, is 106 dB which is 2×10^5.

$$= 4000 \text{ M}\Omega$$

The output resistance in both cases is greatly reduced; to

$$R_o = 75 \Big/ \left(1 + \frac{2 \times 10^5 \times 1000}{100\,000}\right) \Omega$$

for the inverting connection, and

$$R_o = 75 \Big/ \left(1 + \frac{2 \times 10^5 \times 1000}{1000 + 100\,000}\right) \Omega$$

The non-inverting circuit has a slightly higher gain and a slightly lower output resistance, but a much higher input resistance.

for the non-inverting connection. Both these are less than 0.04 Ω.

There are several circuits based on the inverting voltage amplifier which we will see in the next chapter are widely used in circuits for converting between the analogue and digital forms of data. The **summing amplifier** is a simple inverting amplifier with multiple input resistors (Fig. 3.2a). The total current i approaching the virtual earth point at the input to the op-amp is still equal to the current i_2 leaving that point through the feedback resistor. However the current i_1 is now the sum of the currents in each of the input resistors. Thus

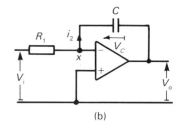

(a)

(b)

Fig. 3.2 (a) The summing amplifier. (b) The integrator

$$i = V_a/R_a + V_b/R_b + V_c/R_c + \ldots$$

and

$$i = -V_o/R_2$$

Since we still have $i_1 = i_2$ then

$$V_o = -[V_a(R_2/R_a) + V_b(R_2/R_b) + V_c(R_2/R_c) + \ldots]$$

This expression shows that the relative effect of each input voltage, V_n, on the overall output voltage is determined partly by its amplitude but also by the scaling factor R_2/R_n.

The **integrator** makes use of a capacitor in the feedback path, so the feedback element is reactive rather than resistive, but we can still deal with the current $i_2(t)$ which is now, in general, a time-varying quantity. Since $i(t) = dq(t)/dt$ and $q(t) = CV(t)$, we can write

$$i_2(t) = C \cdot dV_c(t)/dt$$

But, from Fig. 3.2b, since point X is at virtual earth,

$$V_c = -V_o \text{ so } i_2(t) = -C \cdot dV_o(t)/dt$$

Again using rule 2,

$$i_1(t) = i_2(t)$$

so

$$V_i(t)/R_1 = -C \cdot dV_o(t)/dt$$

or, rearranging and integrating to isolate $V_o(t)$,

$$V_o(t) = \frac{-1}{CR_1} \int V_i(t) dt + V_o(0)$$

The **analogue comparator** is a circuit which compares the magnitudes of two signals and provides a digital true/false-type voltage output indicating which signal is the greater. The output is normally TTL compatible. Essentially, the comparator is an op-amp without negative feedback, and therefore has a very high gain. In use, a small differential signal between the two inputs, typically of the order of a few millivolts, is sufficient to drive the output stage of the amplifier into either saturation or the cutoff state, but the

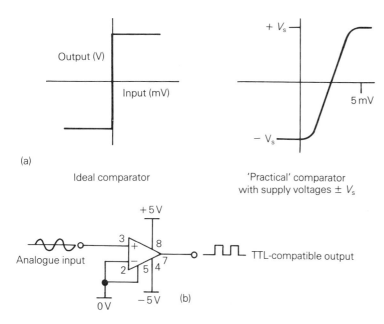

Fig. 3.3 (a) Comparator characteristics. (b) Zero crossing detector using the LM 360 comparator

circuit is designed to withstand these extreme conditions and to recover quickly. These devices are commonly used in applications where small analogue signals are to be detected and digital output signals generated. A zero-crossing detector, for example, detects when the input analogue signal changes from negative polarity to positive. The circuit shown in Fig. 3.3(b) generates a TTL-compatible output signal and switches to the positive level when the input analogue signal reaches only a few millivolts above zero. There are many comparators to choose from, each with its own particular advantages. The National Semiconductor LM361, for example, can recover from overdrive variations from 5 mV to 500 mV in less than 3 ns; the LM211 operates from a single +5 V supply and its open-collector output circuit can switch currents as large as 50 mA at up to 50 V.

In order to measure ac signals, conversion to dc is usually necessary and diode rectification is an obvious approach. But with low-level signals there are problems when the signal is of the same order as, or less than, the voltage drop across the diode. In these circumstances a half-wave **precision rectifier** can be used, in which the diode is placed in the feedback loop of an op-amp. The amplifier output voltage goes sufficiently positive to overcome the forward voltage drop of the diode, which, in effect, is divided by the open-loop gain of the op-amp. The simple circuit shown in Fig. 3.4(a) is limited in practice because the feedback loop is effectively broken during negative voltage excursions, and this can cause the op-amp to go into saturation. To overcome this effect two diodes are used, as shown in Fig. 3.4(b). During a negative excursion, diode D_2 conducts, giving negative

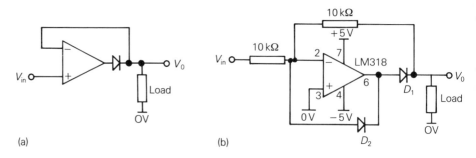

Fig. 3.4 (a) The precision rectifier. (b) A practical example

feedback, and the virtual ground at the input ensures that the op-amp output voltage goes no more negative than the diode voltage drop. Throughout this period diode D_1 is reverse biased and the output is at virtual ground or less, depending upon the loading. During the positive excursion, diode D_1 conducts, and the output voltage follows the input.

The practical operational amplifier

We have so far assumed that the op-amp is ideal in all respects. Although modern commercial op-amps approach the ideal in terms of gain and input and output resistance, they do suffer from imbalances in the internal circuitry which can lead to problems in practice. In a practical integrator circuit, for example, it is common to shunt the integrating capacitor with a high-value resistor to prevent the amplifier input bias current from charging the capacitor and to eliminate the long-term integration of any input offset voltage; for slow changes in the signal the capacitor is then maintained in the discharged state. However, the penalty paid is that the integrator will not operate at frequencies below about $1/2\pi RC$, where R is the resistor value. If low-frequency operation is necessary then an op-amp with a very low offset voltage and negligible input bias current, such as the National Semiconductor LF411A, must be used. Further improvements can be made by increasing the value of the integrating capacitor and lowering the value of the resistor R_1. Note that the integrating capacitor must have a low-leakage dielectric such as polystyrene or teflon.

Many transducers produce signals of only a few millivolts and, in many applications, the signal must be carried to the amplifier over relatively long lengths of wire. Noise signals in the form of spurious and unwanted voltages, perhaps mains hum, picked up along the wire can easily swamp the signal, and differences in earth potential at different points in the system can further aggravate matters. The problem can be overcome to a great extent by using a difference amplifier fed by two wires, closely twisted together, often inside a common earthed screen, and by taking care with the earthing points. Any noise signals picked up in this arrangement are common mode signals, and the difference amplifier is designed to amplify only the differential mode

Such wires are called 'twisted pairs'.

See Horrocks again.

signals. However, the difference amplifier has zero common mode gain only if the signal paths from inverting and non-inverting inputs are identical, and, in practice, that cannot be achieved.

The output voltage is, in fact,

$$V_o = A_1 V_1 - A_2 V_2$$

where A_1 is the amplification along the non-inverting path, and A_2 is the amplification along the inverting path.

By rearranging the equation to a sum and difference form

$$V_o = \frac{A_1 + A_2}{2} \times (V_1 - V_2) + \frac{A_1 - A_2}{2} \times (V_1 + V_2)$$

If $A_1 = A_2$ then $V_0 = A_{dm}(V_1 - V_2)$, as expected.

In the first term, $(A_1 + A_2)/2$ represents the differential mode voltage gain, A_{dm}, and is the average of the two gains, A_1 and A_2. The second term relates to the common mode voltage, V_{cm}, and can be interpreted as the average of the two input voltages multiplied by a factor $(A_1 - A_2)$. This factor is the unwanted common mode gain, A_{cm}. The common mode noise voltages introduced into a system can already be much larger than the signal voltages, and the common mode gain must be kept to an absolute minimum in order to generate as near a perfect differential mode signal at the output as possible. The relative ability of a difference amplifier to maximize the differential gain at the expense of the common mode gain is the common mode rejection ratio which is defined as

$$\text{CMRR} = |A_{dm}|/|A_{cm}|$$

A value greater than 90 dB is normally expected.

It is normally quoted in dB.

A pressure transducer develops an output voltage of 40 mV and is connected to a difference amplifier which has a differential mode gain of 46 dB and a common mode rejection ration of 80 dB. Find the maximum common mode noise signal that can be tolerated if the component of output voltage caused by the noise signal is not to exceed 1% of the required signal.

Worked Example 3.2

The output voltage

$$V_o = A_{dm} V_i + A_{cm} V_n$$

$A_{dm} = 46$ dB, which is equivalent to a gain of 200, giving

$$A_{dm} V_i = 200 \times 40 \text{ mV} = 8 \text{ V}$$

CMRR = 80 dB, equivalent to 10 000, and

$$A_{cm} = A_{dm}/\text{CMRR} = 200/10\,000 = 0.02$$

$(A_{cm} V_n)$ must not exceed 1% of 8 V, which means that the maximum noise voltage,

$$V_{n(max)} = 0.08/0.02 = 4 \text{ V}$$

Other practical limitations in the performance of an op-amp arise from the needs of the manufacturer in designing and producing the circuit. The main

Even a small offset voltage at an early stage in the op-amp can cause the output voltage to settle at the maximum positive or negative value. Another useful reference here is *Op-amps*, 2nd edn.

$$R = R_1R_2/(R_1 + R_2)$$

problems for the user are offset voltages and bias currents at the amplifier inputs, drift of the output voltage over extended periods of time, and bandwidth and slew-rate limitations. For correct operation the op-amp circuitry must be perfectly symmetrical and an offset voltage occurs when it is not possible to retain absolute symmetry. The offset voltage manifests itself as a non-zero output voltage even when the two inputs are shorted together and connected to earth. The offset effect is greatly reduced by use of negative feedback, but, where the residual effect cannot be neglected, external compensation must also be used.

Bias currents occur at the amplifier inputs and are unavoidable where bipolar transistors are used in the input stages. These currents generate bias voltages across the external resistances used in the circuit and again upset the symmetry. The effect can be neutralized by generating an equal voltage at each input, thus converting the unwanted voltages to common mode.

For the detection and simplification of very small voltages or currents, op-amps with a very high input resistance and correspondingly low bias currents are required. Such amplifiers use a combination of bipolar and FET technology to achieve input bias currents which are measured in picoamps (10^{-12} A). If advantage is to be taken of the high input impedance in use, it is often necessary to employ guard rings around the input pins on the printed circuit board to reduce leakage currents. We return to guard rings later in the chapter.

Drift is a problem with all dc-coupled amplifiers since it is impossible to differentiate between slowly changing signal voltages, which may be of importance, and slowly varying internal voltages caused by temperature changes, power supply fluctuations, component ageing and so on, which should be rejected. Drift has the same effect as a changing offset voltage and is quoted as $\mu V/°C$ for temperature changes, $\mu V/V$ for power supply fluctuations and $\mu V/month$ for general ageing.

In a good quality amplifier we expect a high gain for differential mode signals and a high CMRR, together with high input resistance, low output resistance and negligible input offset voltage which is virtually unaffected by changes in temperature and power supply voltages. The gain should be accurately controlled over a wide range of operating conditions. Some single op-amps with negative feedback can approach this ideal specification in many details, but, when extra demands are present, it is often necessary to turn to one of the special amplifiers which have been developed. For example, in order to obtain a good gain from an inverting amplifier, the ratio R_2/R_1 must be large and that implies a low value for R_1, but reducing R_1 also reduces the input resistance and reduces the CMRR. This limitation can be overcome by using an *instrumentation amplifier* consisting of several op-amps.

The conventional three op-amp instrumentation amplifier, Fig. 3.5(a) is a difference amplifier, with gain determined by R_1 and R_2 as usual, preceded by input buffer amplifiers which enhance the CMRR. R_4 is adjustable to trim for maximum dc CMRR. The R_5-C network is sometimes included to allow adjustment for minimum output voltage when a 10 kHz 2 V pk-pk common mode signal is applied. Each input to the difference amplifier has a non-inverting buffer amplifier with heavy negative feedback, so that the high input resistance is maintained. The adjustable resistor, R_A, gives fine adjustment of

(a)

Inverting input

Non-inverting input

Input buffers

Difference amplifier

R_1 R_2 R_3 R_4 R_5 R_A R_{B1} R_{B2} C V_o

(b)

AM427 instrumentation amplifier

Forward dc gain	120 dB min
Input offset voltage	25 μV max
Offset voltage drift	0.6 μV/°C max
Slew rate	2.8V/μs
Input resistance	1.5 MΩ

(c)

$R_1 = R_3 = 500\Omega \pm 0.1\%$
$R_2 = 20\text{k}\Omega \pm 0.1\%$
$R_4 = 19.8\text{k}\Omega \pm 500\Omega$
      potentiometer
$R_5 = 100\text{k}\Omega$
      potentiometer
$C = 1000$ pF
$R_{B1} = R_{B2} = 5\Omega \pm 0.1\%$
$R_A = 390\text{k}\Omega + 100\Omega$
      potentiometer

Fig. 3.5 The instrumentation amplifier: (a) circuit diagram; (b) typical performance figures; (c) component values

the gain of the buffer stages and hence controls the overall gain of the amplifier very accurately. A practical instrumentation amplifier would use a low-noise op-amp such as the Intersil AM427 which has the characteristics summarized in Fig. 3.5(b). When used with the values listed in Fig. 3.5(c) the amplifier has an accurately controlled overall gain of 60 dB and, if balanced source resistances are used, a CMRR of more than 100 dB.

Chopper stabilization

Where the elimination of drift is essential, chopper or other specialized amplifiers are necessary. The **chopper-stabilized amplifier**, Fig. 3.6, achieves very low drift characteristics by separating the signal into two portions: the high-frequency part being amplified by an ac coupled amplifier, and the low-frequency components, which are blocked by a series capacitor, being treated separately. This latter part, which contains the drift signal, is converted to an ac form by means of a chopper circuit, amplified by an ac amplifier, demodulated to regain its dc form and coupled back into the amplifier. By this means the returned signal counteracts the dc offset and drift of the main amplifier by a factor equal to the gain of the ac amplifier,

69

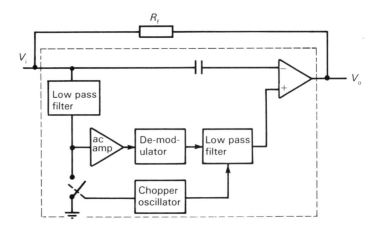

Fig. 3.6 The chopper-stabilized amplifier

Low-pass filters are covered later in the chapter

and the ac amplifier, of course, introduces no offset effects. The frequency response curve for the chopper-stabilized amplifier is non-uniform since the dc component of any signal is amplified by two amplifiers in series, but in practice this is unimportant as the overall gain is controlled by the external feedback arrangements. The drift values achieved in commercial chopper-stabilized amplifiers are very low indeed. For example, the amplifier quoted previously, the AM227, when used with chopper stabilization, achieves a drift as low as $2 \mu V/year$!

A further refinement of chopper stabilization is used in the commutating auto-zeroing (CAZ) amplifier. This device is designed specifically for low-frequency applications, typically from dc to 10 Hz, and extremely low long-term drift characteristics can be achieved. The principle of operation can be understood with reference to Fig. 3.7. The amplifier commutates, or switches,

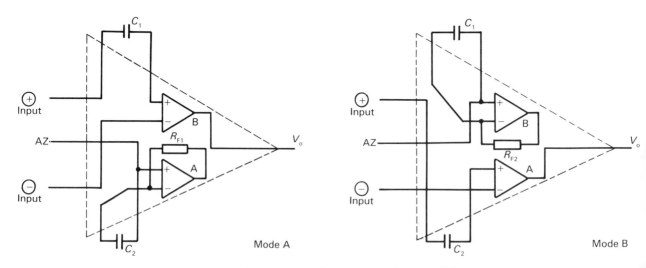

Fig. 3.7 The commutating auto-zeroing amplifier

between two modes A and B, and a signal on the autozero line, AZ, defines the datum to which the amplifiers are zeroed. In mode A, op-amp A is connected as a unity gain amplifier and charges C_2 to a voltage equal to the dc input offset voltage and noise signals. In mode B this voltage is connected in series with the input in such a way as to cancel out the input offset and noise voltage. At the same time, capacitor C_1 is being charged to the dc input offset voltage and noise signals of amplifier B. This switching between modes carries on continuously under internal control. The result is an amplifier with a long-term drift measured in tenths of a microvolt per year, but one drawback is the need to restrict operation to frequencies well below the commutation frequency. A typical application is the digital readout torque wrench which uses the Intersil ICL7606 amplifier. In conjunction with a special chip providing both analogue-to-digital conversion and digital readout capability, a very compact circuit is achieved. An ideal amplifier would be able to cope with rapidly changing input signals as well as slowly varying signals, and amplify each component of the signal by the same amount. In practice, the range of frequencies which can be handled successfully is limited by shunt capacitance which is inherent in the amplifier and surrounding circuitry. For practical reasons, the gain of a typical op-amp without feedback is designed to fall off at a frequency of only 10 Hz or so, and continues to fall at 6 dB per octave as the frequency increases. At higher frequencies, other capacitances begin to have an effect and the gain falls more rapidly. Amplifier bandwidth is defined in terms of the 3 dB point, which is the frequency at which the gain has fallen to $1/\sqrt{2}$ of its original value. It is then said to be '3 dB down'. With negative feedback applied, the low-frequency gain is reduced by a factor $(1 + A\beta)$, but the 3 dB point occurs at a frequency higher than the original by the same factor. Figure 3.8 shows that the bandwidth with negative feedback applied, is $B(1 + A\beta)$, where B is the

The RS7600 amplifier specification lists 0.2 μV/year.

See Chapter 4 for details of analogue-to-digital conversion.

6 dB/octave = 20 dB/decade.

A similar point occurs at very low frequencies

This point is also known as the 'half-power point'. β is the feedback fraction.

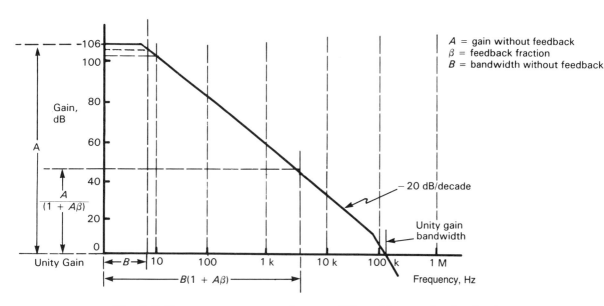

Fig. 3.8 Frequency response of an amplifier with negative feedback

bandwidth without feedback. Thus a practical circuit with heavy negative feedback can have a nominal bandwidth much greater than that indicated by the op-amp itself. An alternative definition of bandwidth is the frequency at which the forward gain of the amplifier has fallen to one. The *unity gain bandwidth* is independent of feedback and is normally several hundred kilohertz or higher. However, in interpreting quoted bandwidth figures it must be remembered that a further limitation is present, because, although the circuitry in modern op-amps operates at very high speed, there is a limit to the rate at which it can respond to input changes. This applies in particular to changes causing large output swings, typically in excess of 1 V. The maximum rate of change of the output voltage, in response to a step change of signal at the input, is known as the **slew rate** of the amplifier. This is defined as the rate of change of output voltage under large signal conditions, and can be determined by applying a high-frequency squarewave to the amplifier input and measuring the time taken for the output to change. Slew rate has dimensions of volts per second, but is more meaningful if quoted in volts per microsecond. The value for the 741 op-amp, for example, is normally quoted as 0.5 V/μs, rather than 500 000 V/s! As the frequency of the input signal increases, the slew-rate limitation means that, for a large amplitude signal, the output tends to become a triangular waveform with a frequency determined by the slew rate. At lower signal amplitudes the slew-rate limitation does not become apparent until rather higher frequencies, but it is always present.

A more useful measure of the bandwidth is therefore often quoted, and is known as the **full-power bandwidth**. This is defined as the maximum frequency of a full voltage sinusoidal input signal which can be delivered at the amplifier output, without slew-rate limiting. Quite often it is found that the response time of a measurement system is limited, not by the response of the transducer, but by the frequency response or slew rate of the signal-conditioning amplifier chosen. Table 3.1 illustrates the wide range of op-amps typically available to the designer.

In any electrical system involving the interfacing of sensitive devices, such as transducers and amplifiers whose response is measured in millivolts and

Table 3.1

	Forward gain dB	Unity gain (small signal bandwidth) MHz	Full-power bandwidh MHz	Slew rate V/μs
AD544*	100+	14	0.2	15
3554†	100+	70	16	1000
741‡	100+	1.3	0.01	0.5

* Analog Devices; FET input.
† Burr-Brown; high-speed, FET input.
‡ Many manufacturers; general purpose.

microamps, noise problems inevitably occur. The most common cause is externally generated electrical interference from ac power circuits, in particular the mains. As we saw in Chapter 1, for example, the magnetic cartridge of a record player is prone to noise induced from the mains transformer.

The mains supply can be responsible for several voltage fluctuations, not only sinusoidal interference signals at 50 Hz. These fluctuations are generally caused by current surges as heavy power circuits switch on and off in the vicinity of the electronic circuitry, though fluorescent lighting can also create interference. Transient voltage 'spikes' can extend to hundreds of volts in magnitude, and contain frequency components in the range between 100 kHz and 10 MHz. Anyone who has had the misfortune to use sensitive mains-powered equipment in a room near to a lift will many times have had just cause to request a change in environment! Electromagnetic radiation arising from current surges produced by arc welding equipment, or the heavy-duty motors operating intermittently, such as in lifts and in pumping equipment, cause havoc with instrumentation systems.

When building up an instrumentation system it is always prudent to assume that electrical noise will be present. It is often thought that because equipment is 'earthed' or 'grounded' it is immune to external electrical interference, but this is not so. The mains supply to any equipment involves three lines: 'line', 'neutral' and 'earth'. The line and neutral connect to the primary of the power transformer and the earth is connected to the metal frame housing the equipment (Fig. 3.9). Interference from these lines can be classified as common mode noise, differential mode noise and radiated noise.
Common mode noise relates to signals which are induced equally into the line and neutral return via the earth conductor, and this occurs particularly when cables are subject to radiation interference. Now, the conductor that serves as the system earth can be a piece of wire, or a printed circuit track, or the metal shielding case itself, and it does not have zero resistance. In fact at high frequencies the resistance can be surprisingly large because of the skin effect. If currents are flowing in any part of the conductor, then circuits earthed at different points cannot be assumed to be at the same common potential (Fig. 3.10a). To be safe, it is necessary to provide a common earthing point

As the frequency increases so the current becomes confined more and more to the surface of the conductor. This increases the effective resistance and is known as the skin effect. At 50 Hz the effect is negligible, but in the megahertz range and above it must be taken into account. The increase in resistance is proportional to $\sqrt{f}$.

Fig. 3.9 Methods by which electrical noise enters a screened unit

(a)

$$V_i = V_o - I_n R$$

(b)

(c) Leakage resistance

Fig. 3.10 Noise problems with earth loops: (a) the input signal at amplifier 2 does not equal the output signal at amplifier 1; (b) a common earthing point ensures that $V_i = V_o$; (c) use of a guard ring minimizes stray resistance and capacitance noise problems

for all sensitive circuits, as in Fig. 3.10(b). The noise currents shown can be generated either internally or externally; the effect is the same and can lead to faulty operation. Where connection to a common point is not possible, and the circuit involved is particularly sensitive, as, for instance, in the use of remote thermocouples, an isolation amplifier may be used to provide a 'floating' input, or output. No ground connection is required since the system is isolated and free from noise interference.

Differential noise is induced around the line-neutral loop but is usually suppressed by the capacitive filters which are used on the dc power supplies. Line-borne *radiated* noise, however, can be very difficult to remove. It is noise radiated into the system by the power lines together acting as an aerial, and only great care in screening at the design stage will ensure freedom from trouble. Screened transformer windings (Fig. 3.9) are essential to help minimize radiated noise from the primary, and the use of screened cables for the signal lines is most important.

The simplest way to remove a great deal of mains noise is to incorporate mains filter units as a matter of course. These consist of low-pass filters (Fig. 3.11) in which, ideally, the inductors are wound on separate cores.

Fig. 3.11 A typical mains filter

However, in order to ensure that the line current flowing in the circuit does not cause magnetic saturation, such cores have to be quite large. In order to reduce the size and, therefore, the cost of the filters, most commercial units have the windings sharing the same core and wound so as to give magnetic flux cancellation. These filters are effective against common mode interference but not against differential mode noise. Typical quoted noise figures, such as 30 to 40 dB noise suppression over a frequency range up to 30 MHz, must, therefore, be treated cautiously if differential noise problems are suspected.

In addition to external noise interference, a common source of trouble when using a sensitive op-amp arises from the effects of stray ac and dc currents at the amplifier input. This is a particular problem with the very high input resistance of modern op-amps where input bias current of a few picoamps can cause problems. A useful technique in such cases is to provide 'guard rings', which are conductive paths placed close to the sensitive parts of the circuit (Fig. 3.10(c)). The ring is maintained at the same potential as the point it is protecting, but provides a low-resistance path for the stray currents. Since the ring and the protected point are at the same potential no current flows between them. The guard also serves to reduce stray capacitance effects of nearby circuitry, the capacitance being to the ring rather than the sensitive area. On a printed circuit board the guard rings will consist of copper track totally encircling the sensitive section with 'jumper' wires used for interconnections.

The minimization of noise by the inclusion of appropriate filters is particularly useful where the frequency range of the noise signal differs greatly from that of the desired signal. The mains hum in the record player, for example, can be reduced by a high-pass filter connected in series with the transducer. Severely attenuating signals at mains frequency and below can eliminate hum and turntable rumble, but the bass response of the transducer to the audio signals is also severely attenuated. Similarly, a low-pass filter which attenuates the higher frequencies can be used to reduce hiss and other surface noises from a record, but again care must be taken not to attenuate the wanted higher audio frequencies. In some applications a very crude low-pass filter, in the form of a large capacitance connected across the amplifier input, can be effective. The strain gauge load cell and instrumentation amplifier system for weighing vehicles (see Fig. 2.11) is unlikely to generate any signal components at a frequency higher than a few hertz. If the impedance of the bridge is typical, at about $120 \, \Omega$, then a $500 \, \mu F$ capacitor strapped across the input terminals of the amplifier will effectively attenuate all but the wanted, slowly varying signals (though not any common-mode noise signals).

Passive filters inevitably introduce some attenuation at all frequencies and if this is not acceptable active filters must be used. By cascading high-pass and low-pass filters a bandpass filter can be constructed. The response can be adjusted to pass any required band of frequencies with severe attenuation of frequencies above and below the band. Conversely, a band-rejection filter can be designed to attenuate only the required band. Such *notch filters* are used to attenuate the signal components in a narrow band centred on the mains frequency when used for mains interference protection, and similar filters

The high-pass filter; 6 dB/octave fall-off

The low-pass filter; 6 dB/octave fall-off

Effect of filter action on a square wave

High-pass active filter: with the values shown, $f_L = 1$ kHz, $A = -1$, fall-off = 12 dB/octave

Low-pass active filter: with the values shown, $f_H = 1$ kHz, $A = -1$ fall-off = 12 dB/octave

protect the 'in-band' signalling frequencies used in the telephone network from interference from subscribers' speech signals. Further discussion of filters is beyond the scope of this book and interested readers are referred to more specialized texts, such as *Electrical Instrumentation and Measurement Systems* by B. A. Gregory.

If noise is present within the signal frequency range then very little can be done to remove it. Truly random noise in a repetitive signal can often be suppressed by an averaging process which can give a worthwhile improvement in signal-to-noise ratio. In some cases the noise is a direct consequence of the operation of some device within the system and is thus predictable. Such *synchronous* noise can be minimized by the use of a sample-and-hold circuit arranged to 'hold' just before the onset of the expected noise signal and to return to the tracking mode when the noise has decayed sufficiently. This technique is limited and works well only when a slowly varying signal is affected by noise voltages of a transient nature.

Composite characteristics (i); the band-pass filter with band-width, $B = f_H - f_L$

Composite characteristics (ii); the band-rejection (notch) filter with bandwidth $B = f_L - f_H$

Modulation

The conversion of dc and very low frequency signals to an ac form to facilitate amplification without the problem of drift, which is the basis of the chopper amplifier, is also used extensively in telemetry systems where signals from outlying transducers may have to be sent over considerable distances. These modulated ac carrier systems also provide good protection from interference of the type mentioned in the previous section, such as power circuits (which generate mains hum at 50 Hz), fluorescent lighting (giving hum at twice the mains frequency), and welding equipment, arc furnaces and dc load switching in general (with voltages at frequencies in the MHz range).

Although we may often refer to sensor outputs as 'dc signals' they are, of course, time varying, though very slowly. The output signal from many transducers, notably bridge circuits, is the product of a dc supply voltage, V_s, and the electrical changes, $x(t)$, introduced by the transducer in response to the parameter variations. In general terms,

$$V_o(t) = KV_s \, x(t)$$

If the signal was genuinely dc, i.e. non-varying, we would not need to measure it continuously and life would be much simpler.

The function $x(t)$ is normally a complex periodic function but we can conveniently represent it as a Fourier series:

$$x(t) = x_0 + \sum_{n=1}^{\infty} x_n \cos[2\pi f_n t + \phi n]$$

Considering only the variation of the signal from its dc value, and simplifying our phase reference to give $\phi_n = 0$, we can simplify the function to

$$x(t) = \sum_{n=1}^{m} x_n \cos 2\pi f_n t$$

The limits are shown as 1 to m since the slowly varying signal contains very few harmonics, and f_m is a few hertz at most.

If we change the supply voltage to an ac voltage at a constant frequency (of a few kilohertz), V_s becomes

$$V_s(t) = \hat{V}_s \cos 2\pi f_s t$$

Modulation is a multiplicative process so the output voltage now becomes

$$V_0(t) = K\hat{V}_s \sum_{n=1}^{m} x_n \cos 2\pi f_n t \cdot \cos 2\pi f_s t$$

This means that for each component of the transducer input, x_i, we have an output voltage

$$V_c(t) = K\hat{V}_s \, x_i \cos 2\pi f_i t \cdot \cos 2\pi f_s t$$

$$= \frac{K\hat{V}_s x_i}{2} [\cos 2\pi (f_s + f_i)t + \cos 2\pi (f_s - f_i)t]$$

$\cos(A+B) + \cos(A-B) =$
$2\cos A \cos B$

Thus, if the highest frequency component in our signal is, for example, 4 Hz, and our supply voltage is at 1 kHz, the modulated output voltage has components between 996 Hz and 1004 Hz. This modulated signal can now be amplified in a conventional ac-coupled amplifier, and we can ensure that most unwanted signals are rejected if we restrict the bandwidth of the amplifier to the narrow range of acceptable frequencies. Mains voltages at 50 Hz, for instance, will give modulated components at 950 Hz and 1050 Hz and it can be arranged that these fall outside the passband.

Note that modulation only protects from drift occurring in circuits after the modulator.

The demodulation process, which is necessary to recover the original signal, is also a multiplicative process. The modulated signal, $V_0(t)$, is again multiplied by the modulating signal, $V_s(t)$, to give for each component

$$x'(t) = \hat{V}_s \cos 2\pi f_s t \frac{K\hat{V}_s x_i}{2} [\cos 2\pi(f_s + f_i)t + \cos \pi(f_s - f_i)t]$$

$$= K\hat{V}_s^2 x_i \cos 2\pi f_i t \cos^2 2\pi f_s t$$

$2\cos A = 1 + \cos 2A.$

$$= \frac{K\hat{V}_s^2 x_i}{2} \cos 2\pi f_i t [1 + \cos 2\pi \cdot 2 f_s t]$$

$$= K'\hat{V}_s x_i \cos 2\pi f_i t + K'\hat{V}_s x_i \cos 2\pi f_i t \cdot \cos 2\pi \cdot 2 f_s t$$

where $K' = K\hat{V}_s/2$.

By use of a low-pass filter to reject frequencies above f_m, the second term in this expression is removed and the original signal is retrieved to give

This is a form of amplitude modulation fully discussed in *Telecommunications Principles* Chapter 3.

$$x'(t) = K'\hat{V}_s \sum_{n=1}^{m} x_n \cos 2\pi f_n t$$

An **isolation amplifier** is another special-purpose amplifier which is used when very small differential signals are carried on top of large common mode voltages. Great care is taken to maximize CMRR. The Datel-Intersil AM227, for example, has a quoted CMRR of typically 176 dB, though, as is to be expected, to achieve this level of performance the circuit layout is critical. This type of amplifier consists of an input stage which is isolated from the output stage so that it can safely handle the very high common mode voltages. Traditional isolation amplifiers have used transformer coupling of an amplitude-modulated high-frequency carrier, but opto-isolation is becoming more common. Since an isolation amplifier is suitable for the amplification of low-level, low-frequency signals in the presence of high common mode interference, it is also useful in remote data acquistion where the amplifier must, of necessity, be some distance from the transducer.

In an amplitude-modulation system the amplitude of the supply voltage is varied to reflect the instantaneous value of the transducer signal, but the frequency of the resulting wave is constant. An alternative modulation technique is preferable under certain circumstances, particularly when transducers based on capacitance or inductance changes are used. The basic circuit is an oscillator whose frequency is governed by some form of tuned circuit, dependent on the value of capacitance or inductance. As the transducer output varies, so the frequency of the oscillator varies, giving a form of frequency modulation (FM). In general terms the sinusoidal voltage generated by the oscillator is given by

Variation in the term $[2\pi f_c t + \phi_c]$ is called 'angle modulation'; if f_c is varied we get frequency modulation, and if ϕ_c is varied we get phase modulation.

$$V_c(t) = \hat{V}_c \cos[2\pi f_c t + \phi_c]$$

If we cause the instantaneous value of frequency, f_n, to vary in sympathy with the transducer signal $x(t)$, where

$$x(t) = \sum_{n=1}^{m} x_n \cos 2\pi f_n t$$

we have

$$V_c(t) = \hat{V}_c \cos[2\pi(f_c + K_c x(t))t + \phi_c]$$

where K_c is a modulation constant which determines the maximum frequency deviation of the modulated signal from the unmodulated carrier frequency. Again choosing our phase reference to give $\phi_c = 0$

$$V_c(t) = \hat{V}_c \cos[2\pi(f_c + K_c x_i \cos 2\pi f_i).t]$$

For detailed analysis see standard texts such as *Digital and Analog Communication Systems*, Chapter 6.

for each component of the transducer signal. This waveform is a complex function which we need not consider further here, and it is sufficient to recognize that, although the amplitude of the oscillator voltage is constant, the instantaneous frequency indicates the transducer signal amplitude at that instant. One convenient method of retrieving the original signal is to convert the oscillator output to squarewave form and use a digital counter to measure the pulse repetition frequency.

The analogue multiplexer or scanner

In many instrumentation applications we need to process a number of different analogue signals in order to provide the required control function. The measurement of pressure, for example, may have to be accompanied by the measurement of temperature; or perhaps a data acquisition system for an internal combustion engine may need to monitor torque, fuel flow, temperature, water pressure and many other parameters. In such cases it is desirable to be able to utilize a single processor and merely to switch each input to it in turn. This is the action of the multiplexer.

A multiplexer contains multiple bilateral analogue switches which share a common output, as in Fig. 3.12. An on-chip address decoder selects the appropriate input, indicated by the applied binary code. The multiplexer switch is essentially a field-effect device with a low ON resistance, typically 25–300 Ω. Multiplexers can be made from CMOS transmission gates, but many are fabricated with both bipolar and field-effect transistors on the same chip. This is the Bi-FET process in which the bipolar devices provide the decoding and switch driving circuitry, so ensuring compatability with standard logic circuits using TTL and CMOS technology, and the FET switch, when ON, maintaining an almost constant resistance over a wide range of input voltages.

The fact that the ON resistance of the switch is not zero must be taken into account when considering the accuracy of any system. The error caused by the finite ON resistance of the switch in Fig. 3.13 is given by

$$E\% = 100/[1 + R_{in}/(R_{on} + R_s)]$$

where R_{on} is the ON resistance of the switch, R_{in} is the input resistance of the following stage and R_s is the source resistance of the input signal.

Strictly this is known as 'time division multiplex' (TDM).

A typical 8-channel multiplexer, the National Semiconductor LF13508, has an ON resistance of 380 Ω and can switch a signal in the range $\pm$ 11 V.

Fig. 3.12 An 8-line-to-1-line multiplexer (MUX)

79

Fig. 3.13 Effect on accuracy of finite ON resistance of the switch

It is possible to ignore the leakage current when the switches are OFF since their resistance is then very high and the current is usually of the order of nanoamps.

The settling time is defined as the time taken for the output signal to reach the desired level, to within a predetermined accuracy, and is dependent upon the time constant

|| indicates 'in parallel with'. The CR time constant is clearly explained in *Transistor Circuit Techniques*, Appendix 2.

$$C_s[R_{in}\|(R_{on} + R_s)]$$

where C_s is the sum of multiplexer output capacitance and all the stray capacitance associated with the output. Typically the output capacitance of a multiplexer can be measured in tens of picofarads. When using these devices it is important to keep the source resistance as low as possible and the load resistance, R_{in}, as high as possible.

Worked Example 3.3

With an ON resistance of 400 Ω and a source resistance of 1 kΩ, the percentage error generated when a buffer amplifier with an input resistance of 10 MΩ is used is

$$E\% = 100/[1 + 10 \times 10^6/400 + 1000)]$$

$$= 0.014\%$$

If the total capacitance associated with the output is 20 pF, the output time-constant is 28 ns, implying that, given a voltage step input, the output will reach 63% of its final value in 28 ns.

Summary

The op-amp is the basic building block of circuits used in processing analogue signals. The performance of a circuit is modified by negative feedback and the gain is determined by the feedback components. Practical circuits are not perfect and the performance is affected, to a greater or lesser extent, by drift, slew rate, bias currents and voltages, and common mode amplification. Specialized circuits have been developed to deal with very slowly varying signals (using chopper stabilization), and signals occurring as small variations in a large common mode signal (using isolation amplifiers). Modulation methods are used in the remote sensing of variables in order to provide protection from interference. When a number of different analogue signals are to be processed, it is usually convenient to deal with each in turn by means of a multiplexer.

Review questions

1. In what ways do practical op-amps fall short of the ideal?
2. What is indicated by a high value of CMRR?
3. Explain the action of a chopper-stabilized amplifier.
4. What factors limit the bandwidth of an amplifier, and what effect does the application of negative feedback have on the bandwidth?
5. Suggest some possible causes of electrical noise.
6. What noise effects can the isolation amplifier help to minimize?
7. Why is modulation employed in telemetry systems?
8. Explain the operation of an analogue multiplexer.

Further reading

1. *Op-amps*, G. B. Clayton, Butterworth, 2nd edn, 1979.
2. *Feedback Circuits and Op-amps*, D. H. Horrocks, Van Nostrand, 1983.
3. *Telecommunications Principles*, J. J. O'Reilly, Van Nostrand, 1984.
4. *Digital and Analog Communication Systems*, K. S. Shanmugam, Wiley, 1979.
5. *Transistor Circuit Techniques*, G. J. Ritchie, Van Nostrand, 2nd edn, 1983.

Problems

3.1 The op-amp of Fig. 3.1(a) has a feedback resistor of 470 kΩ and an input resistor of 46 kΩ. What is the voltage gain?

3.2 In Fig. 3.1(b) what value should R_1 be for a voltage gain of 20 if $R_2 = 100$ kΩ?

3.3 An op-amp has a gain of 10^5 and a CMRR of 80 dB. The input from a bridge circuit is a difference signal of 10 μV with a common mode signal of 500 μV. What are the common mode voltage gain, the output voltage and the error caused by the common mode signal?

3.4 An integrator has a feedback capacitor of 1 μF and an input resistor of 1 MΩ. If a steady potential of +4 V is applied to the input, what value will the output voltage reach after two seconds?

3.5 A multiplexer analogue switch has an 'on' resistance of 200 Ω. If the signal source sampled had a resistance of 2 kΩ, and the buffer amplifier is as shown in Fig. 3.1(a) with $R_1 = 10$ kΩ, what value should R_2 be in order to counteract the loading errors caused by the switch and the op-amp input resistance?

3.6 If the multiplexer in question 3.5 has an output capacitance of 25 pF, how long would it take the output of the amplifier to reach 4 V when the input signal changes suddenly from 0 to 6 V?

4
Signal conversion

Objectives

☐ To introduce techniques for
    digital-to-analogue conversion
    analogue-to-digital conversion.
☐ To explain the need for sample-and-hold methods and the operation of a
typical circuit.
☐ To consider conversion between voltage and frequency, and conversion
using synchros.

In previous chapters we have seen how transducers are used to relate
real-world signals and their electrical analogues. Frequently, in order that
such signals can be generated or accepted by digital systems, data converters
are required. These are known as digital-to-analogue coverters (DAC) and
analogue-to-digital converters (ADC).

The digital-to-analogue converter

The task of converting digital signals to their analogue equivalent is fairly
straightforward. Digital systems, in general, use voltages to represent the
binary values; a positive voltage, commonly somewhere between 3 and 15 V,
to represent a '1', and approximately 0 V to represent '0'. The most popular
digital-to-analogue conversion method uses a resistance ladder network in
conjunction with electronic switches and an op-amp as represented in Fig. 4.1.
There are as many switches as bits in the word to be converted, and each can
apply a voltage to the $2R$ resistor of $+V_{ref}$ when closed, and 0 V when open.

Fig. 4.1 Digital-to-analogue conversion using the $R - 2R$ ladder network

The ladder is terminated by $2R$ to ground at one end and $2R$ to virtual ground at the amplifier, and, since the reference voltage generator has negligible resistance to ground, the impedance seen at any node is, therefore

$$R_0 = R + 2R\|(R + 2R)\|(R + \ldots 2R\|2R)$$
$$= 2R$$

If we assume that the most significant bit (msb) is at '1' and all other bits are at '0', so that switch S_1 is closed and all other switches are open, the voltage at node 1 is $V_{ref}/3$. The amplifier has a gain of $-3R/2R$, so the corresponding output voltage, V_o, is given by

$$V_o = -V_{ref} \times \frac{R}{3R} \times \frac{3R}{2R}$$
$$= -V_{ref}/2$$

If switch S_2 is closed instead of S_1, the voltage at node 2 is $V_{ref}/3$, leading to a voltage of $0.5V_{ref}/3$ at node 1. By extension, each time we move to the switch next furthest from the amplifier, the resulting voltage is halved. In fact, the voltage at node 1 contributed by any switch, S_n, is $V_{ref}/(3 \times 2^{n-1})$. The superposition theorem shows that for any combination of switch positions the output voltage will be given by

$$V_o = V_{ref}(B_1/2 + B_2/4 + B_3/8 + \ldots B_n/2^n)$$

where $B_1 \ldots B_n$ have the binary values '1' or '0' corresponding to the data word presented to the switches.

The great advantage of this type of circuit is that it is not the absolute values of the ladder resistors that are important, but their ratios. Ladder networks for digital-to-analogue converters are readily available, constructed from thick or thin film circuits suitable for hybrid integrated circuit fabrication. Laser trimming of the resistors is used when very high accuracy is required. Practical converters based on the $R - 2R$ ladder often contain built-in registers to hold the binary data, and the Ferranti ZN425E (Fig. 4.2)

The superposition theorem is discussed in standard texts such as *Basic Electrical Engineering*.

Fig. 4.2 The Ferranit ZN425E DAC

is a typical example. This device contains its own (optional) reference voltage supply and an internal counter, which can supply the binary code to the ladder switches. This facility, as we shall see later, is of use when ADCs are constructed.

<table>
<tr><td>Worked example 4.1</td><td>

What is the resolution of an 8-bit DAC? If a ladder network is to be used with a voltage reference source of 2.5 V, to what accuracy should the source be maintained and what is then the resolution in volts?

The resolution of the converter is determined by the weighting of the least significant bit, which in this case is 1/256. The resolution would therefore be quoted as one part in 256. This represents an accuracy of $1/256 \times 100\%$, i.e. 0.4%. The output of an $R - 2R$ ladder is $V_{\text{ref}}/(3 \times 2^{n-1})$. For the least significant bit, then, the output signal is $2.5/(3 \times 2^7)$, i.e. 6.5 mV. Since we are dealing with a system accuracy of 0.4% it would not make sense to provide a reference voltage with an accuracy maintainable to only, say, ±10%. On the other hand, it is not necessary to provide a costly device with more stability than is needed. In this case, a reference with an output guaranteed to remain within ±0.1% would be reasonable.
</td></tr>
</table>

Digital-to-analogue conversion is also possible using binary weighted resistor values at the input of a summing amplifier (Fig. 4.3). The closing of the switches, shown diagramatically in the figure, is controlled by the binary signals. The input to the op-amp is a current summing node and the output voltage is proportional to the sum of the currents into the node. Thus, for the values shown

See the previous chapter, Fig. 3.2. With a negative reference voltage and an inverting amplifier, V_o is positive.

$$V_o = -V_{\text{ref}}\left(S_1 \times \frac{5 \times 10^3}{10 \times 10^3} + S_2 \times \frac{5 \times 10^3}{20 \times 10^3} + S_3 \times \frac{5 \times 10^3}{40 \times 10^3} + \ldots \right.$$
$$\left. S_8 \times \frac{5 \times 10^3}{1.28 \times 10^6}\right)$$
$$= -V_{\text{ref}}(S_1/2 + S_2/4 + S_3/8 + \ldots S_8/256)$$

The 'all ones' digital code, 11111111, therefore, indicates the output voltage is

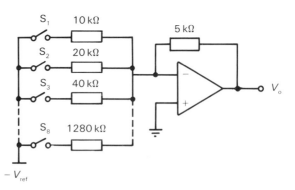

Fig. 4.3 Digital-to-analogue conversion by current summation

$$V_o = 255V_{ref}/256$$

$$= 0.996V_{ref}$$

and the minimum output from the converter, 00000001, corresponds to a voltage

$$V_o = V_{ref}/256$$

$$= 0.004V_{ref}$$

The accuracy and stability of this type of converter is critically dependent upon the absolute accuracy of the resistors and their temperature stability. Such devices are available commercially, and typical of the range is the National Semiconductor DAC1200, a 12-bit converter using thin film resistors and an internal 10.24 V reference source for binary operation. This device has a minimum output voltage step of 10.24/4096 or 2.5 mV. The full-scale output voltage is

$$V_o = V_{ref}(1/2 + 1/4 + 1/8 + \ldots + 1/4096)$$

$$= 0.9997566V_{ref}$$

$$= 10.2375 \text{ V}$$

The DAC1200 is mounted in a 24-pin dual-in-line package, and, with its optional internal amplifier connection, can provide either a voltage output or the summation current.

Two important parameters must be considered in the specification of a digital-to-analogue converter: **monotonicity** and **linearity**. The output of the converter is said to be monotonic if, at every point in its range, a change in the least significant bit (lsb) of the input code causes a step change in the output voltage in the same direction. That is, a non-negative output step should be produced for an increasing input step. Errors in the bit-weighting of the resistors can lead to non-monotonicity.

Linearity is measured, either as a percentage of full-scale output or as a fraction of the lsb, as the maximum amount by which any point on the transfer characteristic deviates from the ideal straight line passing through zero and the full-scale output, as in Fig. 4.4(a). It is important to remember

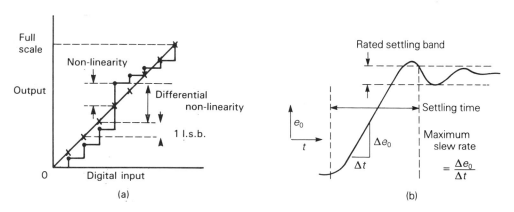

Fig. 4.4 Response of a DAC: (a) non-linearity; (b) response to an input step of 1 lsb

that the resolution of a converter is simply the number of bit inputs provided, indicating the smallest analogue increment that the converter can produce. **Differential non-linearity** specified as a fraction of the lsb, is the maximum difference between the actual and ideal size of any one lsb analogue increment. Note that, if this non-linearity becomes more negative than one lsb, the converter becomes non-monotonic. An n-bit converter exhibiting say -1.5 lsb of differential non-linearity could be made monotonic, though with a resolution reduced to $n - 1$ bits, by holding the least significant digital input bit permanently at '0'. This approach can provide significant cost savings in certain applications, where n-bit resolution is not warranted.

An increasingly common use for DACs is in the provision of grey scales or colour shades for computer-generated CRT displays. Here the digital output from the computer must be converted into an analogue voltage to modulate the beam current in the CRT, and the main problem is one of timing. For a standard TV tube, the time taken for the beam to scan a single line is 62.5 μs. If we assume a good resolution picture with, for example, 256 pixels per line, the time for each pixel is 62.5/256, i.e. 240 ns. For each spot to be modulated with, say, 16 grey scales, or colour intensities, the 4-bit converter must be capable of responding in a time of 20–30 ns. The fastest rate at which code conversions can take place in a converter is termed the **through-put rate**, and throughput rate = 1/(settling time). The output of a converter is considered to have settled to its new equilibrium value when it enters, and stays in, a specified band. This is illustrated in the diagram of Fig. 4.4(b). The main factors limiting the speed are the large value resistors needed in the higher resolution converters, which give rise to large time constants, and the switching speed of the internal transistor switches. To reduce the former problem, weighted current sources are sometimes used.

The analogue-to-digital converter

Perhaps the simplest form of ADC is the single-ramp ADC. The conversion relies on the comparison of the analogue input signal with an accurately timed ramp signal. A control unit initiates the ramp voltage and gates clock pulses into a counter. The output from the comparator, Fig. 4.5, changes from '0' to '1' when the ramp voltage exceeds the analogue input voltage, and the value then stored in the counter is proportional to the applied signal. The conversion time for this arrangement depends upon both the clock frequency and the magnitude of the unknown analogue voltage since, for an input voltage near to the maximum, the binary counter will have to count almost to its maximum value. Small changes in clock frequency and non-linearities in the ramp signal can lead to errors in the conversion accuracy. A low-cost single-ramp ADC, can be constructed around the Texas Instruments TL507C, which requires only an external oscillator.

A more efficient and more commonly used conversion technique is known as the **successive approximation**, or **put-and-take**, method, in which a DAC is used in conjunction with a register to generate a voltage which is compared with the unknown voltage. The most significant bit of the register is first set to '1'. If the resulting analogue voltage, V_r in Fig. 4.6, is greater than the

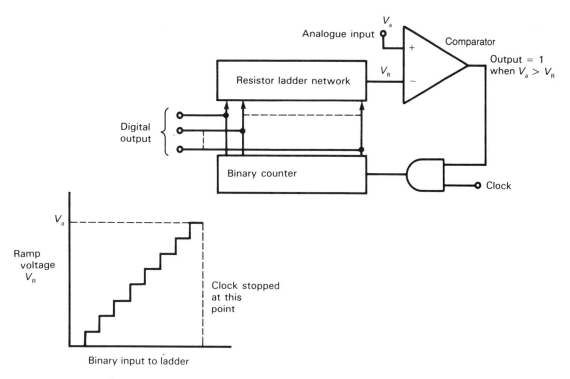

Fig. 4.5 Single-ramp ADC. The clock is stopped when $V_R = V_a$

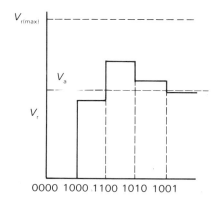

Fig. 4.6 Successive approximation analogue-to-digital conversion

unknown voltage, V_a, then the msb is reset to '0' and the next most significant bit is set to '1' instead. If the generated voltage is now less than V_a, that bit is retained at '1' and the next most significant bit is also set to '1'. The process continues until voltages V_r and V_a are equal to within the limits of resolution. The computed value for each bit position is registered during the following clock period so the conversion is always completed in just $(n + 1)$ steps, where n is the resolution or number of bits in the register. Conversion speed is high, with 10-bit quantization in 1 μs being not uncommon. Figure 4.6 shows the output typical of this type of converter, the final

Note that the comparator is usually biased by 1/2 lsb so that it will switch at the 'ideal' point.

The conversion time is constant regardless of the amplitude of the unknown voltage.

For details of three-state operation see Chapter 6.

digital value in this case being 1001. A commonly used converter of this type is the Plessey/Ferranti ZN448E, which is an 8-bit converter with three-state outputs enabling it to be interfaced directly to microcomputer systems. The linearity of the device is quoted as $\pm 1/2$ lsb, and the differential linearity as ± 1 lsb. The conversion time is typically $10\ \mu s$, and a reference voltage generator is provided for use if required.

The **up–down integrator**, or **dual-ramp** converter, enables automatic zeroing capability to be built in, and is shown diagrammatically in Fig. 4.7. After the unknown voltage, V_a, has been applied, switch S_1 is closed. The output from the integrator, V_o, is a positive ramp function and the comparator registers an output which primes the AND gate, allowing the counter to start counting clock pulses. When the most significant bit sets, i.e. the counter registers $1000\ldots0$, the switch control unit changes the input so that the integrator receives the negative reference voltage. V_o now ramps down at a rate defined by V_{ref}, and as long as the comparator output is positive, the counter continues to increment. During this period the output gates are enabled and the count is displayed. As soon as the integrator output becomes negative the comparator switches and stops the count. The time t_1 taken to set the most significant bit of the counter is constant, but the time t_x, needed for the negative-going ramp to reach 0 V, is proportional to the amplitude achieved by the positive-going ramp. In terms of the voltages, and assuming that $V_o = 0$ at $t = 0$, we have

This technique has important noise rejection qualities; if the time t_1 is set to equal the period of the noise signal (20 ms for mains hum, for example), the rejection is infinite.

$$V_o(t_1) = \frac{1}{CR} \int_0^{t_1} V_a\, dt = \frac{V_a/t_1}{CR}$$

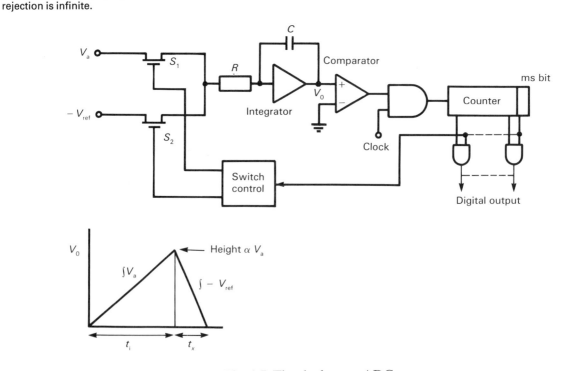

Fig. 4.7 The dual-ramp ADC

Also,

$$V_o(t_x) = \frac{V_{ref}t_x}{CR}$$

Now

$$V_o(t_1) = V_o(t_x)$$

therefore

$$\frac{V_a t_1}{CR} = \frac{V_{ref}t_x}{CR}$$

whence

$$V_a = V_{ref}t_x/t_1$$

Since V_{ref} and t_1 are known constants, t_x is proportional to the applied voltage V_a, and the count accumulated during the time t_x is, therefore, a digital representation of V_a. The accuracy of conversion is independent of the clock frequency and hence of any drift associated with the clock frequency. For increased resolution it is necessary only to increase the capacity of the counter, though this, of course, increases the conversion time. For example, the Harris ICL7109 is a 12-bit dual-ramp converter and can operate at a rate of up to 30 conversions per second. This type of converter is commonly used in the digital voltmeter where its high resolution more than compensates for the relative slowness of operation. To the user, 20 to 30 conversions per second can give the impression of instantaneous readout!

The conversion time can be an important parameter, and although a quoted conversion time may, on the face of it, seem fast, a simple calculation will show that one can very soon get into difficulties. Consider, for example, a system which needs to convert a 5 V sinusoidal signal to digital form; we have

$$e = 5 \sin \omega t$$

so the rate of change of the voltage is

$$de/dt = 5\omega \cos \omega t$$

The maximum rate of change occurs when $\sin \omega t = 0$, giving

$$(de/dt)_{max} = 5\omega$$
$$= 10\pi f$$

$d^2e/dt^2 = -5\omega \sin \omega t$, which equals zero when $\sin \omega t$ is zero, i.e. $\cos \omega t = 1$. $\omega = 2\pi f$.

For an 8-bit resolution,

$$1/2 \text{ lsb} = \frac{1}{2} \times \frac{\text{input voltage range}}{2^8} = \frac{10}{512}$$
$$= 19.5 \text{ mV}$$

The Teledyne TSC-8703CJ has a quoted conversion time of 1.25 ms, so, setting $dt \to \Delta t = 1.25$ ms and $de \to \Delta e = 19.5$ mV, we get

$$\left(\frac{\Delta e}{\Delta t}\right)_{max} = \frac{19.5 \times 10^{-3}}{1.25 \times 10^{-3}} = 10\pi f$$

where f is the maximum frequency possible while retaining a conversion error not greater than $1/2$ lsb.

Thus

$$f = \frac{19.5}{12.5\pi} = 0.49 \text{ Hz!}$$

This response appears to be inadequate for almost any application but, as is shown later, sample-and-hold techniques can overcome the limitation.

For very high speed conversions, such as those required in video and radar data conversion and image processing, where the resolution demanded is typically limited to six to eight bits, a parallel conversion method can be employed. For an n-bit output code, $2^n - 1$ analogue comparators are used, and this approach has become possible only because the cost and scale of integration of the circuitry have improved dramatically in recent years. The high conversion speed leads to the name 'flash' being used with this type of converter. Each comparator compares the analogue input voltage, V_i, against a reference voltage which is derived from a series resistor chain, so that each bias point differs from the adjacent point by $V_{ref}/2^n$. Those comparators for which the reference voltage is below V_i saturate and produce a logical '1' signal; those for which the reference voltage is above V_i register a logical '0'. Since all the comparators change state almost simultaneously, the quantization process is very fast, limited only by comparator delay. However, the output parallel code is not in binary so a further conversion stage is required.

Worked example 4.2

Design a 3-bit flash ADC capable of converting voltages in the range 0 to +4 V at a rate of 10^7 conversions per second.

A 3-bit converter requires $2^3 - 1$ comparators, and the resolution is $V_{ref}/2^n$. In this case we will set $V_{ref} = +4$ V, so the resolution becomes 0.5 V. Each comparator must be biased to the midpoint of the bit value, as in Fig. 4.8(a) where $V = +4$ V. The absolute value of R is not critical and depends on the load capabilities of the reference supply and the current drain through the comparators, though this is usually negligible. We will use a value for R of 1 kΩ. A suitable comparator is the LM360, which has TTL-compatible outputs, a response time of less than 20 ns and an input bias current of 5 μA. The circuit layout is shown in Fig. 4.8(b). The +4 V reference is derived from a Ferranti ZNREF040 reference diode. The output codes and the corresponding binary codes are as follows:

C_7	C_6	C_5	C_4	C_3	C_2	C_1	A	B	C
0	0	0	0	0	0	0	0	0	0
0	0	0	0	0	0	1	0	0	1
0	0	0	0	0	1	1	0	1	0
0	0	0	0	1	1	1	0	1	1
0	0	0	1	1	1	1	1	0	0
0	0	1	1	1	1	1	1	0	1
0	1	1	1	1	1	1	1	1	0
1	1	1	1	1	1	1	1	1	1

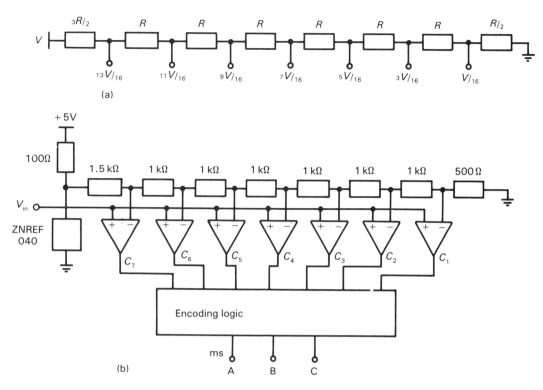

Fig. 4.8 The flash ADC: (a) bias resistor chain; (b) circuit arrangement

The necessary encoding logic can most easily be implemented using a priority encoder such as the 74LS147. With a propagation delay through the logic of about 20 ns, the overall delay should not exceed 50 ns, which is adequate for performance at 10 MHz.

Flash converters are usually available only for low-resolution applications because of the large number of comparators required. A typical device, the Datel-Intersil ADC302, converts to an 8-bit output at a conversion rate up to 50 MHz.

Sample-and-hold circuits

In most of the analogue-to-digital converters described above, unless the input signal is static there is uncertainty as to exactly when the indicated value matched the input value. If it is necessary to know the value at a specific time, a sample-and-hold (S/H) circuit is used to capture the value and to hold it until the next sample is taken. In other applications, especially when dealing with rapidly changing signals, or with signals from different sources in a data acquisition system, S/H circuits allow sampled values to be held steady until processing is completed. Sample-and-hold circuits are also useful if the low-level analogue signal to be monitored is periodically subject to high-level transient spikes, or glitches, caused by electrically adjacent clock or similar circuits. The use of the S/H circuit to maintain the analogue level during the noise burst protects sensitive amplifiers from overload and subsequent slow recovery.

The basic sampling concept is illustrated in Fig. 4.9(a). When the switch is closed, the voltage across the capacitor follows, or tracks, the applied voltage, V_{in}. When the switch is opened the capacitor retains the voltage attained at that time. Figure 4.9(b) shows a simple practical form of the circuit, in which the amplifiers with unity gain serve to minimize the loading effect of the capacitor. The FET provides an almost perfect switch, having an OFF resistance greater than $10^8 \, \Omega$ with zero offset voltage (unlike a bipolar transistor), and an ON resistance which can be less than $100 \, \Omega$. The hold time is limited by the leakage of charge from the capacitor, and leakage currents through the FET and the op-amp, together with polarization effects in the capacitor dielectric, all contribute to the losses. As a result, the voltage decays, or droops. Many variants of the basic S/H circuit are commercially available and the important parameters governing the choice are acquisition time and droop rate. These are defined with reference to Fig. 4.10. The acquisition time is the time taken by the output voltage to reach its final value, within a specified error band, and depends upon the maximum charging current the input amplifier can supply to the capacitor. The droop rate is the rate of change of the output voltage when the device is in the hold

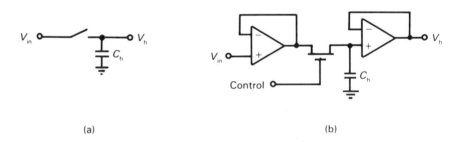

(a)   (b)

Fig. 4.9 Sample-and-hold circuit: (a) the basic concept; (b) a simple practical form

In taking a sample there is an uncertainty as to when the actual value existed. This is the window or aperture time. In an ADC it extends over the entire conversion time but in a S/H circuit it is the signal averaging time during the sample to hold transition.

Fig. 4.10 Definition of parameters

mode, and is a function of the leakage, or droop, current, I_d, and the capacitance, C_h, given by

$$\frac{dV_h}{dt} = \frac{i_d}{C_h}$$

The size of C_h is determined by the required droop rate, usually at the expense of acquisition time.

A typical S/H device, the National Semiconductor LF398, with a capacitor of 1000 pF, has a droop current of 30 pA and an acquisition time to within 0.1% of 4 μs.

The maximum charging current of the Precision Monolithics SMP-10 S/H circuit is 50 mA and the droop current is 100 pA. What is the expected droop rate if the acquisition time for a 3 V step is 6 μs?

Worked example 4.3

The voltage developed across C_h is given as

$$V_h = \frac{1}{C_h} \int i_c \, dt$$

where i_c is the amplifier charging current.

Assuming a constant current, we obtain

A fair approximation in practice.

$$\begin{aligned} C_h &= i_c t / V_h \\ &= (50 \times 10^{-3} \times 6 \times 10^{-6})/3 \\ &= 0.001 \ \mu\text{F} \end{aligned}$$

The droop rate, dV_h/dt, can be obtained by assuming a constant droop current, i_d, and

Again a reasonable assumption since the change in voltage across the capacitor will necessarily be small.

$$\begin{aligned} \text{droop rate} &= i_d / C_h \\ &= 100 \times 10^{-12} / 100 \times 10^{-9} \\ &= 1 \ \text{mV/s} \end{aligned}$$

Voltage-to-frequency conversion

The techniques so far described are direct ADC methods but, in many cases, the conversion of an analogue voltage to a periodic wave is a useful intermediate stage. The integrating digital voltmeter, for example, converts a dc signal into a periodic wave of proportional frequency. The cycles of this wave are then counted during a precisely timed interval, and the resulting count is displayed as the digital representation of the voltage. Another use is in the recording of dc voltages; here conversion to a pulse train whose frequency is proportional to the voltage enables conventional magnetic tape recorders to be used.

This is another form of frequency modulation.

The basic arrangement of the voltage-to-frequency (V/F) converter is shown in Fig. 4.11. When a voltage is applied to the integrating op-amp, a ramp output voltage is generated with a slope proportional to the applied voltage. This ramp is applied to a monostable pulse generator, which produces a pulse of accurately defined width when the ramp input voltage

Fig. 4.11 The voltage-to-frequency converter

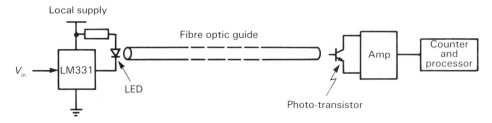

Fig. 4.12 Measurement of voltage with total electrical isolation

reaches a precisely determined level. The pulse is also fed back to a FET switch which discharges the integrating capacitor, thus terminating the ramp. At the end of the output pulse the ramp recommences and the repetition rate of the pulses is proportional to the applied voltage, V_{in}.

V/F converters also find extensive use in cases where voltages are to be measured in high noise environments, or where for other reasons isolation is desirable. Figure 4.12 shows an arrangement using a National Semiconductor LM331 V/F converter with optical coupling to give total electrical isolation.

Frequency-to-voltage (F/V) conversion is also of use, in particular in speedometers or tachometers, where the rotation of a shaft generates a pulse train with a frequency proportional to the angular velocity. Conversion of this frequency to a voltage allows the velocity to be read directly from a meter. The converters are essentially charge pump devices, in which a capacitor is charged from a constant-current source switched on for the period of the incoming pulse, and then discharged from another, smaller, current source during the intervening period. The voltage developed across the capacitor has a maximum amplitude proportional to the repetition rate of the incoming pulses, and, after filtering, a steady dc output voltage is produced.

Synchro-to-digital conversion

A **synchro-to-digital converter** (SDC) converts three-wire position information from a synchro to digital form with commercial units providing a resolution between 10 and 18 bits. Two common types of SDC are available with each intended for specific types of application. These are the tracking converter and the successive approximation converter.

The tracking SDC takes signals from the synchro and converts them to twin signal form at a lower voltage level, more suitable for the processing circuitry. The signals are processed independently as sine and cosine channels and pass through multiplying DACs (MDACs) with sine and cosine law response respectively. Assume that the present digitally indicated position is φ; the transformer output voltages, Fig. 4.13, are

$$V_1 = V \sin \omega t \sin \theta$$

and

$$V_2 = V \sin \omega t \cos \theta$$

where θ is the synchro angle.

The input signals to the error amplifier are then

$$V_1' = V \sin \omega t \sin \theta \cos \varphi$$

and

$$V_2' = V \sin \omega t \cos \theta \sin \varphi$$

so that the error signal is

$$V_e = V \sin \omega t (\sin \theta \cos \varphi - \cos \theta \sin \varphi)$$

$$= V \sin \omega t \sin (\theta - \varphi)$$

After effective demodulation by the phase-sensitive detector, a dc version of the error signal is produced which is proportional to $(\theta - \varphi)$. This is integrated and used to control a voltage-controlled oscillator (VCO), so that pulses are sent to the counter and, by feedback action, the error is reduced to zero. That is

This two-signal format is the same as that produced by a resolver, so this method can also be used in a resolver-to-digital converter (RDC).

This transformer connection is known as 'Scott connected' and, when fed with voltages at 120° phase difference, gives voltages at the output which are 90° out of phase.

Fig. 4.13 Synchro-to-digital converter: tracking converter

$$\sin(\theta - \varphi) = 0$$

and

$$\theta = \varphi$$

Thus, for a 10-bit converter, the value is updated every $360/1024 = 0.35°$ change in input angle.

The counter, therefore, always takes the digital value corresponding to the synchro angle, θ, and automatically tracks the input with an increment occurring whenever the input angle changes by more than the equivalent of the least significant bit of the counter. The successive approximation converter also operates on resolver format signals and these are converted to digital form by means of conventional successive approximation ADCs. The signals are sampled and held under control of the reference voltage and give the two dc signals

$$V_1 = V \sin \theta$$

and

$$V_2 = V \cos \theta$$

from which the difference angle, $\sin(\theta - \varphi)$, is derived and, as before, φ is adjusted to give $(\theta - \varphi)$ equal to zero. The basic construction is shown in Fig. 4.14.

The successive approximation SDC is very cost-effective when a large number of channels of synchro information must be handled, but the need for the sample-and-hold circuit means that the converted value may be 'out of date'. For example, in a 50 Hz system, the actual conversion may take only 100 μs, but a sample is taken only every cycle of the reference voltage and the maximum delay can be up to 20 ms. The tracking SDC has a much faster response and is also much more tolerant to noise and related unwanted signals. Its operation relies on the ratio between signals, rather than their absolute values, so long lines and the voltage drops they introduce do not cause problems. Furthermore, the use of the integrator smooths out the effects of noise spikes.

The block diagram of Fig. 4.15 shows the feedback arrangements of a

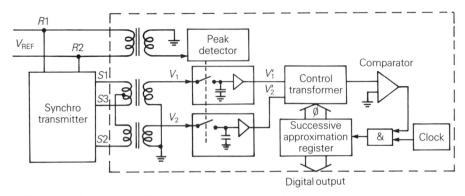

Fig. 4.14 Synchro-to-digital converter: successive approximation converter

Fig. 4.15 Feedback control of servo system

system of the type which would be used to control a large servo system by microprocessor. The synchro senses the position of the main shaft and a tracking SDC (such as the Memory Devices SDC1704) converts the signals to digital form. The error signal is generated within the processor and is used in the feedback loop to control the drive to the motor which rotates the main shaft. The dc velocity voltage involved in the frequency shaping process is the signal from the SDC integrator used to control the VCO.

This example is developed in detail in *Synchro and Resolver Conversion.*

The phase lock loop

The problems of recovering wanted signals from a very noisy background is common to most areas of electronic engineering and a circuit arrangement which has been found to be very useful in 'cleaning up' a signal is the phase lock loop (PLL). As its name implies, this circuit has the ability to lock on to and follow, or track, the phase (or frequency) of a signal and has therefore found wide application in areas as diverse as am/fm demodulation, frequency multiplication and synthesis, and motor control. It can also be used to track and synchronize with a signal which varies in frequency with time, so is useful in voltage-to-frequency conversion and in synchronization of pulse trains. PLLs are available as integrated circuits from many manufacturers, a particularly popular type being the CMOS 4046.

The essential elements of a PLL are shown in Fig. 4.16(a). The phase detector, or comparator, generates an output voltage which is proportional to the phase difference between the input signal, v_{in}, and the reference voltage v_{vco}. This is filtered and fed to a voltage-controlled oscillator (VCO). The phase detector may consist simply of an exclusive-OR gate or can be a more complex edge-controlled circuit. The first type – type I – provides a digital error signal which, when passed through a low-pass filter, gives a dc voltage with an amplitude determined by the relative phasing of the signals, reaching a maximum when they are 180° out of phase.

A second type of detector – type II – generates pulses which are triggered by the positive-going edges of the incoming signals; if the reference signal

(a) Block diagram

(b) The simple phase-detector

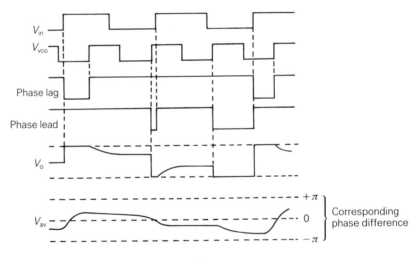

(c) Edge-controlled phase detector waveforms

Fig. 4.16 The phase lock loop

from the VCO leads the input signal a phase lead pulse is generated, and if it lags a phase lag is indicated. No pulses are generated at all if the two signals are in phase, i.e. locked. This is a big advantage over the first type of detector, which can give a 'ripple' or form of phase modulation at the output under these conditions. After passing through a low-pass filter (LPF), the signal amplitude is again proportional to the relative phasing of the input signals and this is independent of the mark–space ratio, unlike the first type.

The output of the LPF is used to control the frequency of the VCO and can be considered as the error signal if we think in terms of a typical feedback control system. The frequency of the oscillator output voltage, which forms the reference signal at the detector, is therefore adjusted in such a way as to reduce the error, i.e. the phase difference, towards zero, and the VCO locks on to the phase of the incoming signal.

With appropriate phase detectors the PLL will operate equally well with sinusoidal input signals. Let us assume that the loop is not yet quite locked in phase but the two signals, v_{in} and v_{vco}, are very close in frequency. The output of the phase detector, v_o, is then a very low frequency beat signal which passes happily through the LPF and causes the VCO frequency to change so as to reduce the error. This eventually results in phase locking when the two frequencies become equal.

We can show the operation of the PLL in general terms if we consider the input signal to be a sinusoid of frequency ω_0; the VCO output is of the same frequency but has a relative phase, $\varphi(t)$, which the feedback signals control. Let

$$v_{in} = \sqrt{2}V_{in} \sin \omega_0 t$$

and

$$v_{vco} = \sqrt{2}V_{vco} \cos[\omega_0 t - \varphi(t)]$$

Use of the cosine makes the maths easier and the additional phase angle is absorbed in φ.

The phase detector acts as a multiplier and gives

$$v_0 = v_{in}.v_{vco}$$
$$= 2.V_{in}.V_{vco}.\sin \omega_0 t . \cos[\omega_0 t - \varphi(t)]$$
$$= V_{in}.V_{vco}\{\sin \varphi(t) + \sin[2\omega_0 t - \varphi(t)]\}$$

$2 \sin A \cos B = \sin(A - B) + \sin(A + B)$.

This consists of an ac term, $\sin[2\omega_0 t - \varphi(t)]$, which is removed by the following LPF, and a dc term which is a simple function of the relative phase, $\varphi(t)$. Thus the signal reaching the VCO is

$$v_{av} = V_{in}.V_{vco}.\sin \varphi(t)$$

The frequency of the VCO output is controlled by the input voltage and, since frequency is the rate of change of phase angle, we can relate input and output of the VCO by

$$\frac{d\varphi(t)}{dt} = K.v_{av}$$

from which $\varphi(t) = K \int v_{av} \, dt$.

Thus the VCO acts as an integrator and introduces a 90° lagging phase shift in the feedback loop. In order to maintain stability the phase characteristics of the LPF, which introduces further phase shift, must be carefully chosen. When suitable component values are used this 'second order' loop arrangement produces a very stable circuit with a precise 'capture range' of frequencies at which phase locking will occur, and good noise rejection properties thanks to the 'flywheel' action of the feedback which smooths out the input fluctuations.

PLLs are widely used in applications ranging from control to communications, and specific detail is given in appropriate textbooks: a good introduction is given in *The Art of Electronics* by Horowitz and Hill. The definitive textbook is *Phaselock Techniques,* 2nd edn by F. M. Gardener.

Summary

This chapter has dealt with DACs and ADCs. The design of such devices is a specialized field of electronics and a wide variety of techniques can be used, the choice being governed by demands of resolution and speed. The general principles of the most popular methods have been introduced, and we have defined important parameters such as monotonicity and linearity. Many ADCs use a digital-to-analogue converter within the conversion loop, others use single or double ramps. However, the successive approximation is probably the most frequently used technique. S/H devices are essential to the accurate conversion of time-varying signals. Finally, V/F and F/V converters and synchro-to-digital converters are seen to be suitable for applications in noisy environments and where rotating machinery needs to be monitored.

Review questions

1. Explain why the operation of the resistor ladder network ADC is critically dependent on the ratio of the relative resistors, and not their absolute values.
2. Define monotonicity and linearity in the context of data conversion.
3. Explain the operation of the flash ADC.
4. Why are S/H circuits necessary in many applications of ADCs?
5. Describe one possible use of a V/F converter.

Further reading

1. *Basic Electrical Engineering*, 5th edn, A. E. Fitzgerald, D. E. Higginbotham and A. Grabel, McGraw-Hill, 1981.
2. *Transducer Interfacing Handbook*, D. H. Sheingold (ed.), Analog Devices, 1981.
3. *Transistor Circuit Techniques*, 2nd edn, G. J. Ritchie, Van Nostrand, 1987.
4. *Data Converters,* G. B. Clayton, Macmillan, 1982.
5. *Data Conversion Products Databook,* Analog Devices, 1988.
6. *Synchro and Resolver Conversion,* G. S. Boyes (ed.), Memory Devices 1980.

Problems

4.1 The converter of Fig. 4.3 is required to give an output voltage in the range 0 to 2.55 V. What value should V_{ref} be?

4.2 The converter of Fig. 4.8b has a reference voltage of 4.00 V. What is the input voltage if the output code is 101?

4.3 A 100 Hz sinusoidal signal with a maximum amplitude 10 V, is to be resolved by an ADC to a resolution better than 0.1%. What aperture time should the S/H circuit have?

4.4 A 12-bit successive approximation converter uses a 2 MHz clock. What is the conversion time?

4.5 The internal DAC of a 4-bit successive approximation ADC generates the following voltages:

bit 0= '1', 0.5 V	bit 1= '1', 1.0 V
bit 2= '1'. 2.0 V	bit 3= '1', 4.0 V

Determine the sequence of register states for applied voltages of (a) 5.8 V, (b) 4.7 V, (c) 3.3 V.

5
Digital processing of signals

Objectives

☐ To introduce the concept of the discrete or digital domain.
☐ To highlight the equivalence of operations in the analogue and digital domains.
☐ To describe the principles of sampling and quantization of an analogue signal.
☐ To consider the performance of basic digital processing circuits and to compare their action with the corresponding analogue circuit.
☐ To review the facilities provided in a typical digital signal processing chip.

Much of the processing necessary in dealing with signals acquired from transducers involves filtering, since the raw signals consist of a large number of components at different frequencies, some of which may interfere with, or mask, the essential components. With analogue, or continuous, signals this can be achieved by the use of op-amp circuits with appropriate feedback components including inductors and capacitors. There are two main types of filter, the first of which allows low-frequency signals to pass but removes all components above a defined frequency. This is the **low-pass filter** (LPF). The second operates in the opposite sense and allows the high-frequency signals to pass but removes the components below a defined frequency. This is the **high-pass filter** (HPF). In theory components within the passband are not attenuated at all and there is a linear relationship between phase shift and frequency, whilst those outside the band are removed completely. As always, in practice these ideals are not met. Two related filter types can be constructed using combinations of low- and high-pass filter sections; when connected in series low-pass and high-pass sections give a **bandpass** filter in which only the band of frequencies lying between the two defined frequencies is transmitted, and when connected in parallel a **bandstop** filter is achieved.

Once we have converted the analogue signals to an equivalent digital representation we can approximate the time or frequency response of a filter by means of digital techniques with shift registers, adders and delay elements. All these features are provided in conventional microprocessors but, in recent years, digital IC devices have been produced in which these particular operations are optimized. Such devices are known as digital signal processors (DSP).

Filtering in the digital domain

A filter is a single-input, single-output circuit in which the output signal, or response, $F(t)$, is a linear function of the input signal, or excitation, $f(t)$. In the analogue domain the input and output signals are both continuous functions of time and are related by a convolution integral

$$F(t) = \int_{\infty}^{-\infty} h(\tau)f(t-\tau)d\tau$$

See *Signals and Systems*, Chapter 2.

where $f(t - \tau)$ indicates a time-shifted version of $f(t)$ and $h(\tau)$ is the impulse response of the filter.

An alternative form of the response is represented in the frequency domain by transforming the time functions using Laplace transforms. Then

$$F(j\omega) = H(j\omega)f(j\omega)$$

The impulse is the response to a unit impulse function (or Dirac function) and completely characterizes the dynamic behaviour of a continuous system.

where $H(j\omega)$ is the frequency response function of the filter and the poles of $H(j\omega)$ give the natural frequencies of the filter. The zeros indicate the frequencies at which the output signal is zero.

In a digital filter the input and output signals are represented by sequences of numbers each with a finite number of bits. The input sequences are derived from the analogue signals by sampling and converting to digital form by use of an ADC. A function, $f(t)$, sampled at regular intervals, T, is written as $f(nT)$ and, since the output words are generated at the same rate, the output function is $F(nT)$. As long as certain minimum conditions are satisfied in sampling the input signal, it is possible to reconstruct a continuous output signal from $F(nT)$ using a DAC. Thus a digital filter can be considered as a specialized form of digital computer, and the flexibility that this introduces means that virtually any relationship between input and output signals can be achieved, not only those that are possible with conventional analogue filters.

The discrete version of the convolutional integral given earlier is

$$F(nT) = \sum_{m=0}^{n} h(mT)\,f(nT - mT)$$

and the discrete equivalent of the impulse function is

$$h(mT) = 0,\ m \neq 0$$

$$h(0) = 1$$

The operation in the digital domain which corresponds to the Laplace transform is the Z transform which, for a sequence $h(mT)$ is defined as

$$H(Z) = \sum_{m=0}^{\infty} h(mT)Z^{-m}$$

and

$$F(Z) = H(Z)f(Z)$$

This is very similar to the frequency response of a system and can be considered as a form of frequency response though, since the output of a digital filter is a series of numbers, this interpretation must be used with care.

The equivalent terms and operations in the analogue and digital domains are summarized in Fig. 5.1.

Analogue Domain	Digital Domain
Continuous signals	Discrete signals
Differential equations	Difference equations
Time domain	Discrete time domain
Frequency domain $(j\omega)$	Z-domain (Z)

Fig. 5.1 Equivalence of analogue and digital domains

Sampling

In order to retain all the time-varying information contained in an analogue signal, when we convert to digital form we must ensure that analogue samples for conversion to digital form are taken sufficiently frequently. A high sampling rate will ensure that all the information is retained and, in the extreme, we could consider the analogue signal itself to be a sampled signal with an infinitely small sampling period. There are advantages, however, in using a lower sampling rate if possible, especially the ability this gives to multiplex several signals so as to time-share a single channel when it is necessary to transmit the data further. Unfortunately, as the sampling rate is reduced, there comes a point at which the ability to interpolate correctly from the samples becomes impaired and this is directly related to the high-frequency content of the waveform. If the sampling rate is too low the reconstituted signal lacks the higher frequency components and at very low rates a false, or aliased, signal is produced (Fig. 5.2).

The limit to which the sampling rate can be reduced without loss of accuracy is defined by Nyquist's sampling theorem which states that, in order to reconstruct a continuous signal containing frequency components up to f_B Hz, the rate of sampling with constant time intervals must exceed a frequency of $2f_B$ samples per second.

> It is safest to create a band-limited signal by means of an anti-aliasing filter before sampling to ensure that unwanted frequencies are not introduced.

The justification of the sampling theorem can be seen diagramatically in Fig. 5.3 which shows the frequency spectra of the original continuous signal and the sampled versions. Sampling of the original band-limited signal lead to an infinitely large bandwidth – in theory containing negative components as well as positive – but the original signal can be recovered by low-pass filtering as long as the individual response curves do not overlap. Overlapping begins when the sampling rate is reduced below twice the highest component in the signal. The low-pass filters available in practice do not have the ideal 'brick-wall' response necessary to allow separation of very closely spaced sections of the spectrum and a sampling rate somewhat higher than that indicated by the sampling theorem must be used, or, alternatively, for a given sampling rate, the signal must be restricted to a narrower bandwidth.

> This is known as the 'Nyquist frequency'. The frequency response of discrete time signals is dealt with in detail in *Signals and Systems*.

Worked example 5.1

The maximum frequency on the telephone network is limited to 3.5 kHz. What is the mininum sampling rate possible for accurate reconstruction of a digital representation of a signal on the network, and what is a more practical rate?

The sampling theorem indicates a minimum sampling rate of 2×3.5 kHz $= 7$ kHz but to offset the use of non-ideal filtering a rate of 8 kHz would normally be used.

> With slowly varying signals from sensors a sampling rate of about 20 Hz is common.

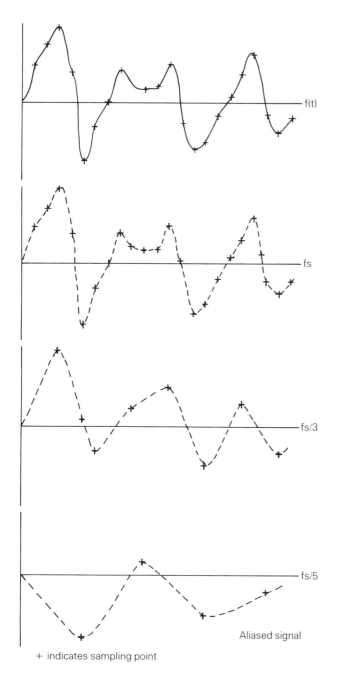

f(t)

fs

fs/3

fs/5

Aliased signal

+ indicates sampling point

Fig. 5.2 Effect of low sampling rate on waveform fidelity

Quantization

Having decided upon a suitable sampling frequency to ensure that the waveform information is retained in full, we must also ensure that the digital

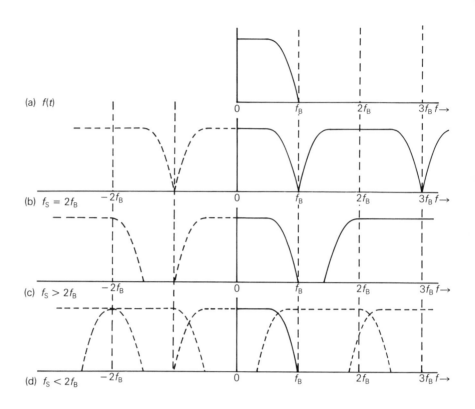

(a) $f(t)$

(b) $f_S = 2f_B$

(c) $f_S > 2f_B$

(d) $f_S < 2f_B$

Fig. 5.3 Spectrum of sampled signal at different sampling rates: (a) original continuous signal spectrum; (b) spectrum of signal sampled at $2f_B$; (c) sampling rate higher than $2f_B$; (d) sampling rate lower than $2f_B$

Sampling and quantization are the essential ideas behind pulse code modulation (PCM). See *Telecommunication Principles* Chapter 6.

Non-uniform spacing can have advantages in offsetting noise effects.

representation of each sample is correct. The process by which the sampled analogue value is allocated a digital equivalent is known as **quantization** and its accuracy is determined by the number of discrete values which are possible between the two limits of the signal. This in turn is governed by the number of bits available in the ADC. For n bits there are 2^n possible values and, if uniformly spaced, the increment for a signal of maximum amplitude $2 V_{pk}$ is $2 V_{pk}/2^n$. (See Fig. 5.4.) Thus the maximum quantization error, that is the difference between the actual signal value and the nearest digital value, is

$$\Delta_{max} = \frac{1}{2}\left[\frac{2V_{pk}}{2^n}\right] = \frac{V_{pk}}{2^n}$$

The quantization error can be considered as a noise signal on top of the analogue input signal and, for a pure sinewave, the noise has an rms value of $\Delta_{max}/\sqrt{3}$. In terms of the signal-to-noise ratio (SNR) this leads to a useful 'rule of thumb' relating SNR to the number of bits in the ADC:

$$SNR = (6n + 1.8) \text{ dB}$$

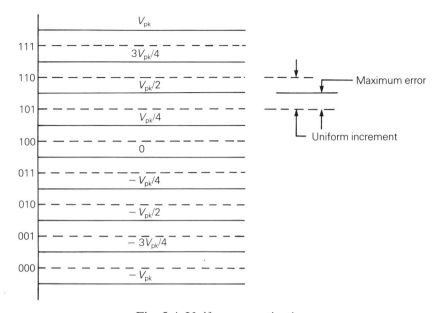

Fig. 5.4 Uniform quantization

where n is the number of bits. The approximate SNR for a range of bit values then becomes:

n bits	SNR dB
6	38
8	50
10	62
12	74
14	86
16	98

An SNR of 70 dB is usually acceptable and it is therefore unnecessary to increase the resolution beyond 12 bits in most cases.

If the quantization resolution of 100 mV peak-to-peak signal on the telephone network is 8 bits what minimum transmission rate, in bits per second, is necessary, and what is the maximum quantization error?

We have seen in Worked Example 5.1 that the practical sampling rate is 8 kHz. The minimum transmission rate is, therefore, 8 bits × 8 kHz giving 64 kbps. The maximum quantization error is

Worked example 5.2

bps = bits per second.

$$\Delta_{max} = \frac{V_{pk}}{2^n} = \frac{(100 \times 10^{-3})/2}{2^8} = 195 \ \mu V$$

Signal averaging

The speed of operation of a computer is very high and, even if the computer is scanning a large number of sensors, it is capable of taking samples from each sensor many times a second. On first thoughts that might seem to be a good thing but, in practice, the sensor signals are inevitably subject to noise disturbances and high rate sampling can make the system over-responsive to the high-frequency components of the noise. High-frequency 'spikes' can be removed by filtering but in many cases an alternative method, making better use of the computer, is to employ signal averaging. In the simplest approach the samples are averaged over a fixed period, giving

$$x_{av} = 1/n \sum_{j=1}^{n} x_j$$

This approach has the disadvantage that the new value is available only at the end of the period so the averaging period must be kept quite short, and, in addition it removes much of any small-scale variation which may be an important feature of the signal. A superior method uses a running average in which it is always the n most recent samples which are averaged. This removes the main noise features but retains much more of the information content, as well as making a new value available at every sampling period.

Linearization of sensor response

The computer can also be used to adjust the received data to compensate for non-linearities in the characteristics of the sensor. A table of pairs of values, acquired in many cases by direct measurement on the sensor, is stored in the computer memory. In operation, whenever a reading is received from the sensor, the computer is programmed to use a 'table look-up' procedure which selects the entry in the table which is nearest to the incoming value and outputs the associated value. We may, for example, be using a thermocouple so the table consists of temperature values paired in each case with the voltage given by the thermocouple at that temperature. Each voltage indication from the thermocouple is used to point into the table and the temperature value is then read directly. Assume the thermocouple operates over the range 0 to 100 °C and provides an output voltage with a resolution of 1% but non-linear. A linear scale can be produced by digitizing the voltage to the required accuracy (7 bits would give 128 different points) and then using this 7-bit data word as the address of a 128 × 7-bit PROM. The value retrieved is the linearized value of the sensor output as shown in Fig. 5.5. High accuracy requires a large number of values and, hence, a large table but, in many cases sufficient accuracy can be achieved by means of linear interpolation between fewer points.

Programmable ROMs are ideal for storage of linearization data where the sensor characteristics do not change appreciably with time. However, electrically alterable PROMs do now allow individual bytes to be erased and reprogrammed and we therefore have the ability to recalibrate a system by adjusting the data in the table if it becomes necessary.

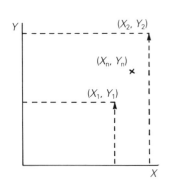

To a good approximation

$$\frac{X_n - X_1}{Y_n - Y_1} = \frac{X_2 - X_1}{Y_2 - Y_1}$$

therefore

$$X_n = \frac{X_2 - X_1}{Y_2 - Y_1}(Y_n - Y_1) + X_1$$

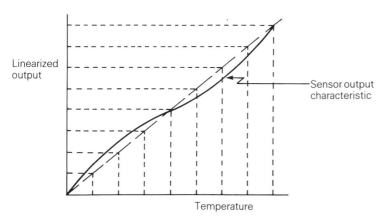

Fig. 5.5 Linearization data

Digital processing circuits

The digital integrator

The form of the integrator is shown in Fig. 5.6(a). The output series, $y(nT)$, is delayed in the shift register and added to the incoming sample, $x(nT)$, so

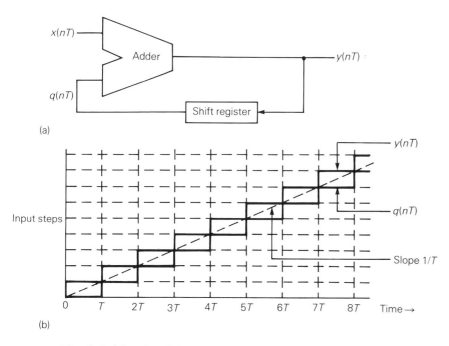

Fig. 5.6 (a) The digital integrator. (b) Digital ramp output

that

$$q(nT) = y(nT - \tau)$$

In effect, q takes the value that y had τ seconds earlier and, in most cases, τ is proportional to $1/f_s$, where f_s is the sampling frequency. The output at any time is, therefore, the sum of the current input value and all previous samples. That is,

$$y(nT) = \sum_{m=0}^{n} x(mT) = x(0) + x(T) + x(2T) + \ldots + x(nT)$$

For a uniform series of step inputs the output of the integrator is a *digital ramp* or *staircase* with an effective 'linear' slope of $1/T$ (Fig. 5.6b).

An analogy can be drawn between the operation of the digital integrator and the analogue integrator we considered in Chapter 3. We saw then that

$$V_o(t) = \frac{-1}{CR} \int V_i(t)dt + V_o(0)$$

and $V_o(0)$ is usually arranged to be zero.

If we convert operations such as differentiation and integration to the frequency domain we find that they reduce to simple arithmetic operations such as multiplication and division and we start by looking at the significance of a time delay which, we recall, gives $f(t - \tau)$ as the form of $f(t)$ when it is time-shifted, or delayed, by τ seconds.

The standard Laplace integral is

$$f(j\omega) = \int_0^\infty e^{-j\omega t} f(t) \, dt$$

from which it follows that

$$f(j\omega + a) = \int_0^\infty e^{-(j\omega + a)t} f(t) \, dt$$

$$= \int_0^\infty e^{-j\omega t} [e^{-at} f(t)] \, dt$$

which is the transform of $e^{-at} f(t)$.

The expression $e^{-j\omega t}$ is known as the **delay operator** and, if $f(j\omega)$ is the transform of the function $f(t)$, then the transform of the delayed function $f(t - \tau)$ is $e^{-j\omega \tau} \cdot f(j\omega)$.

We can express the delay operator in an alternative form by use of the series expansion:

$$e^{j\omega \tau} = 1 + j\omega \tau + \frac{(j\omega \tau)^2}{2!} + \ldots$$

But $j\omega \tau \ll 1$ so

$$e^{j\omega \tau} \sim 1 + j\omega \tau$$

and

$$e^{-j\omega \tau} = 1/(1 + j\omega \tau)$$

Integration of a function in the time domain becomes division by $j\omega$ in the frequency domain, so that

$$\int_0^t f(t)\,\mathrm{d}t \leftrightarrow \frac{f(j\omega)}{j\omega}$$

Assuming, as usual, that $V_o(0) = 0$, the expression for the integrator output becomes

$$V_o(j\omega) = \frac{1}{\tau}\frac{V_i(j\omega)}{j\omega}$$

$$\approx \frac{1 + j\omega\tau}{(1 + j\omega\tau) - 1}\, V_i(j\omega)$$

since $j\omega\tau \ll 1$. Rearranging gives

$$V_i(j\omega) = \frac{1 + j\omega\tau}{1 + j\omega\tau}V_o(j\omega) - \frac{V_o(j\omega)}{1 + j\omega\tau}$$

from which

$$V_o(j\omega) = V_i(j\omega) + \frac{V_o(j\omega)}{1 + j\omega\tau}$$

$$= V_i(j\omega) + V_o(j\omega)\, e^{-j\omega\tau}$$

Comparing this with the discrete form we see that

$$y(nT) = x(nT) + q(nT)$$

where $q(nT)$ is $y(nT - \tau)$, the delayed form of the output, and the two expressions, therefore, have the same basic form.

The effective slope of our staircase waveform is $1/T$, and in the analogue integrator the slope is governed by the *CR* product of the integrator components.

The low-pass filter, LPF

The integrator with additional scaling operations becomes a low-pass filter, and scaling in the digital domain is achieved by shifting and adding. The output of the filter, Fig. 5.7, is

$$y(nT) = K_1\, x(nT) + (1 - K_2)y\{(n - 1)T\}$$

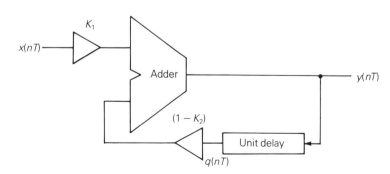

Fig. 5.7 The low-pass filter

and by successive substitution we obtain the series

$$y(nT) = K_1 x(nT) + (1 - K_2)K_1 x\{(n - 1)T\}$$
$$+ (1 - K_2)^2 K_1 x\{(n - 2)T\}$$
$$+ (1 - K_2)^3 K_1 x\{(n - 3)T\} + \dots$$
$$= K_1 \sum_{k=0}^{n} (1 - k_2)^{n-k} x(KT)$$

This represents the time response of the digital filter and $q(nT)$ is related to it by

$$q(nT) = K_1 \sum_{k=0}^{n-1} (1 - K_2)^{n-k-1} x(kT)$$

In comparing the operation of the digital LPF with its analogue equivalent we recall that the output of an integrator is given by

$$V_o(j\omega) = V_i(j\omega) + \frac{V_o(j\omega)}{1 + j\omega T}$$

and, with the scaling factor added, we get the expression for the low-pass filter as

$$V_o(j\omega) = K_1 V_i(j\omega) + (1 - K_2) \frac{V_o(j\omega)}{1 + j\omega T}$$

from which

$$V_o(j\omega) = \frac{K_1 (1 + j\omega T)}{K_2 (1 + j\omega T/K_2)} V_i(j\omega)$$

Also, since $q(j\omega) = V_o(j\omega)/(1 + j\omega T)$

$$q(j\omega) = \frac{K_1}{K_2} \frac{V_i(j\omega)}{(1 + j\omega T/K_2)}$$

For a step input signal

$$q(j\omega) = \frac{1}{j\omega} \frac{K_1}{K_2} \frac{1}{(1 + j\omega T/K_2)}$$

which, in the time domain, becomes

$$q(t) = \frac{K_1}{K_2} (1 - e^{-K_2 t/T})$$

K_1/K_2 gives us the passband gain of the filter, and the 3 dB cutoff frequency is defined as K_2/T radians. The response of the filter to a step input is shown in Fig. 5.8 where the values of K_1 and K_2 are taken to be about 0.5.

The high-pass filter, HPF

The high-pass filter uses a form of differentiator which, in the digital domain, is achieved by subtracting the delay input from the current value of the input, Fig. 5.9(a). The addition of scaling factors modifies the differentiator's action to pass only high frequencies.

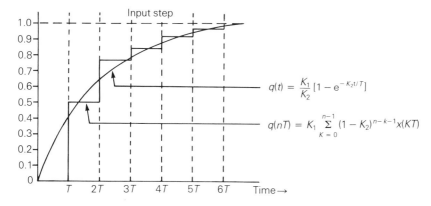

$$q(t) = \frac{K_1}{K_2}[1 - e^{-K_2 t/T}]$$

$$q(nT) = K_1 \sum_{K=0}^{n-1} (1 - K_2)^{n-k-1} x(KT)$$

Fig. 5.8 Response of low-pass filter to step input

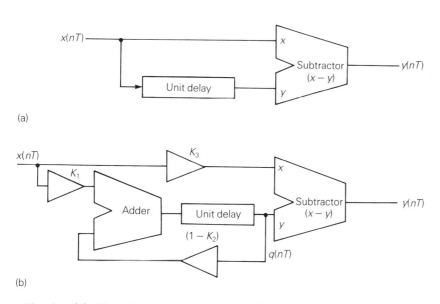

(a)

(b)

Fig. 5.9 (a) The digital differentiator. (b) The digital high-pass filter

The output of the digital differentiator is given by

$$y(nT) = x(nT) - q(nT)$$

where

$$q(nT) = x\{(n-1)T\} \text{ and, again, } q(0) = 0$$

The differentiator's response to a step input is shown in Fig. 5.10(a). When the scaling factors are added, as in Fig. 5.9(b), for $n \geqslant 1$

$$q(nT) = K_1 \sum_{k=0}^{n-1} (1 - K_2)^{n-k-1} x(kT)$$

and

Note that $y(nT) = x(nT)$ when $n = 0$.

$$y(nT) = K_3 x(nT) - K_1 \sum_{k=0}^{n-1} (1 - K_2)^{n-k-1} x(kT)$$

The response of the filter to a step input is shown in Fig. 5.10(b), and again we can compare this with the response in the frequency domain.

In the frequency domain, $q(nT)$ becomes $q(j\omega)$ where

As we derived for the low-pass filter.

$$q(j\omega) = \frac{K_1}{K_2} \frac{x(j\omega)}{1 + j\omega T/K_2}$$

Then

$$y(j\omega) = K_3 x(j\omega) - q(j\omega)$$

$$= K_3 x(j\omega) - \frac{K_1}{K_2} \frac{x(j\omega)}{1 + j\omega T/K_2}$$

$$= x(j\omega) \left[\frac{K_3(1 + j\omega T/K_2) - K_1/K_2}{1 + j\omega T/K_2} \right]$$

If we let $K_3 = K_1/K_2$ this reduces to

(a)

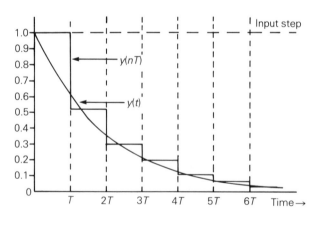

(b)

Fig. 5.10 (a) Differentiator response. (b) High-pass filter response

114

$$y(j\omega) = x(j\omega)\left[\frac{1}{K_2} \frac{j\omega T}{(1 + j\omega T/K_2)} \frac{K_1}{K_2}\right]$$

which is the response of a high-pass filter with a passband gain of K_1/K_2 and 3 dB cutoff frequency at K_2/T.

Transforming, finally, to the time domain:

$$y(t) = \frac{K_1}{K_2} e^{-K_2t/T}$$

More complex filters may be built up from low-pass and high-pass sections and computer-based design aids are available to determine the necessary sampling rates and constants to give the desired gain and frequency response. Sampling rate is always important and the approximations introduced above increase in accuracy as the sampling rate is increased. Even at higher sampling rates care must be taken that difference frequencies involving the sampling frequency and components of the signal do not exist within the passband.

We have seen that a digital filter has a single input stream of data samples and a single output which is the summation of terms involving weighted variants of the input and delayed input values. The inclusion of delayed input terms, so that the output depends on previous outputs, leads to the name *recursive* for this type of filter. A *non-recursive* filter is a special case using only the current inputs and it generates the output simply by weighting the inputs and summing them over a given period.

The expression for the output of a non-recursive digital filter can be given in generalized form as

$$y(nT) = \sum_{k=-\infty}^{\infty} h_k x[(n - k)T]$$

where $x[(n - k)T]$ are the input samples and h_k are the weighting coefficients. In theory the summation is to infinity but in practice, of course, this is not possible and initial conditions must be specified, in effect limiting the summation to the positive terms up to some manageable number defined by the length of the filter.

This is often several hundred stages.

The recursive digital filter produces an output which also takes account of the delayed input values, giving

$$y(nT) = \sum_{k=0}^{m} h_k x[(n - k)T] + \sum_{j=1}^{n} d_j y[(k - j)T]$$

where d_j are the recursive coefficients.

Design of a digital filter to a given specification involves the computation of the best set of coefficients to satisfy that specification and several standard techniques are available. For all practical purposes design is only possible with computer assistance and special software packages have been written to provide the designer with a powerful environment allowing the calculation of the necessary coefficients when the type of filter and required specification are supplied. Figure 5.11(a) gives the set of coefficients prepared by such a program for a simple non-recursive LPF with a cutoff frequency of 50 Hz and sampling at 500 Hz. Figure 5.11(b) is a plot of the expected filter response over 20 points.

See, for example, *Designing Digital Filters*.

Sample points: 512
Sample frequency: 500.00, normalized

Low pass filter, low cut off: 0.20

Rectangular window, length: 20

────────────────── Coefficients ──────────────────

h[0]	=	0.19922	h[1]	=	0.18646	h[2]	=	0.15111
h[3]	=	0.10114	h[4]	=	0.04739	h[5]	=	0.00078
h[6]	= −0.03053		h[7]	= −0.04297		h[8]	= −0.03804	
h[9]	= −0.02139		h[10]	= −0.00078		h[11]	=	0.01635
h[12]	=	0.02493	h[13]	=	0.02347	h[14]	=	0.01395
h[15]	=	0.00078	h[16]	= −0.01102		h[17]	= −0.01749	
h[18]	= −0.01697		h[19]	= −0.01042		h[20]	= −0.00078	

(a)

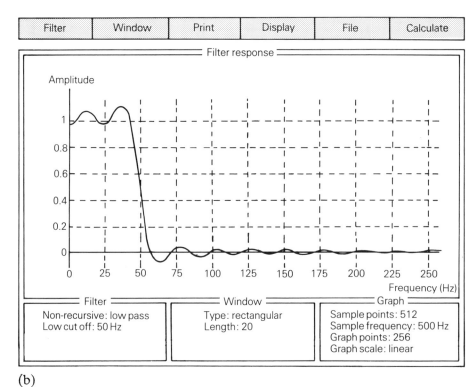

(b)

Fig. 5.11 (a) Non-recursive filter coefficients. (b) Filter response

The digital signal processor

The digital operations of shifting, adding and delaying are all available in conventional microprocessors but the essentially repetitive nature of digital signal processing, and the need for maximum speed to cope with the normally large amount of incoming data, has led to the development of special digital processing chips. The designers of these chips have concentrated on optimizing specific operations by provision of special units such as a multiplier and accumulator unit (MAC), a shifter, and one or more data address generators (DAG). In common with many microprocessors designed for use mainly in industrial control environments a digital signal processor (DSP) is normally based on a type of Harvard architecture rather than the more usual von Neumann architecture. This means that program (or code) memory is kept separate from data memory, though there are facilities for the inclusion of data in program memory if necessary. A higher speed of processing is achieved by pipelining instructions, so that a new operation is begun before the previous one is completed, and by inclusion of dedicated timer-counters to control program looping.

An introduction to the architecture of microprocessors is given in *Computers and Microprocessors*.

Figure 5.12 shows the simplified block diagram of a popular DSP, the Analog Devices ADSP-2100. The processor is built around four buses with an additional internal bus, the result bus, to provide rapid transfer of results between the three main computational sections, which operate in parallel in order to increase throughput. The four buses are:

the program memory address bus, PMA (14 bits)
the program memory data bus, PMD (24 bits)
the data memory adress bus, DMA (14 bits)
the data memory data bus, DMD (16 bits)

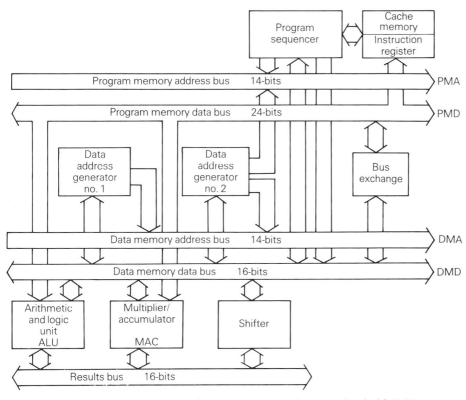

Fig. 5.12 Block diagram of the Analog Devices ADSP-2100 DSP

117

The DMD bus is the primary bidirectional route for both internal and external data and is associated with the DMA bus which carries 14-bit addresses allowing direct addressing of up to 16K 16-bit words. The PMA bus is also 14 bits wide and is used to address up to 16K of external program memory. The bus can effectively be expanded to 15 bits, if necessary, so doubling the size of memory directly addressable. Instructions are returned on the 24-bit PMD bus. In fetching data from the program memory an extra timing cycle is required but, if high-speed operation is absolutely essential, the 16-word cache memory can be used as an alternative source of operands.

The address of the next instruction is generated by the **program sequencer** which works in conjunction with the internal loop counters and fast memory to provide very high speed execution of program loops. Data address generators, DAGs, keep track of up to four **pointers**, or address registers, for data fetch operations, and the provision of two DAGs allows the DSP to deal with two operand fetches simultaneously, again increasing the speed of operation.

The three computational elements are the **arithmetic and logic unit** (ALU) the **multiplier and accumulator unit** (MAC), and the **shifter**. The ALU operates on the 16-bit data words and carries out conventional arithmetic and logic functions such as addition, subtraction, negation and logical AND, OR and exclusive-OR. It also has rudimentry division capability. The shifter is essentially a 16×32-bit *barrel shifter* with exponent detection and comparison circuitry, and extra logic to pass the result either directly to the output register or to OR with the value already in the register. A barrel shifter can perform both logical and arithmetic shifts and place the 16-bit input word anywhere in the 32-bit output register. Both the ALU and the shifter contain a duplicated set of working registers to give fast context switching when dealing with any of the four possible external interrupts.

The MAC implements the high-speed multiplication of two 16-bit words and the 32-bit resultant is routed to a 40-bit accumulator, though only 16- or 32-bit operands are transferred out. Multiplexed inputs and outputs give a wide range of possible source and destination registers.

The arrangements for the use of an ADSP-2100 processor in a basic system are shown in Fig. 5.13.

Summary

The advent of cheap, powerful, high-speed digital processing has encouraged the development of digital techniques to handle all aspects of signal processing. In converting to the digital domain, however, sampling and quantization limitations must be borne in mind. Operations in the analogue domain have direct equivalents in the digital, or discrete, domain, and we have seen how simple low- and high-pass filters can be created. More complex filters can then be synthesized from these basic building blocks. Specialized DSP chips are now available from a range of manufacturers and are optimized to carry out the types of operation commonly encountered in digital signal processing at maximum speed.

Fig. 5.13 Basic DSP configuration

Review questions

5.1 What are the essential elements of a filter and how does a digital filter differ from an analogue filter?

5.2 What advantages can be gained by using a sampled form of data rather than the analogue signal itself?

5.3 Describe the errors which can be introduced by the sampling of a signal and subsequent quantization.

5.4 The signal from a pressure transducer is to be sampled and converted to digital form but it is known that 50 Hz mains cables will lie near the transducer lines and noise signals may be present. What sampling frequency would be appropriate to minimize the noise interaction and what other precautions should be taken?

5.5 What is the approximate signal-to-noise ratio for the system outlined in Worked Example 5.2?

5.6 Describe the special processing units you would expect to find in a DSP.

Further reading

1. *Signals and Systems*, M. L. Meade and C. R. Dillon, Van Nostrand, 1986.

2. *Telecommunication Principles*, J. J. O'Reilly, Van Nostrand, 1984
3. *Instrumentation, Measurement and Feedback*, B. E. Jones, McGraw-Hill, 1977.
4. *Computers and Microprocessors*, 2nd edn, A. C. Downton, Van Nostrand, 1988.
5. *Fundamentals of Modern Digital Systems*, 2nd edn, B. R. Bannister and D. G. Whitehead, Macmillan, 1987.
6. *Digital Signal Processing Design,* A. Bateman and W. Yates, Pitman, 1988.
7. *DSP Products Databook*, Analog Devices, 1989.
8. *Designing Digital Filters*, C. S. Williams, Prentice-Hall, 1986.
9. *Digital Filter Design Handbook*, F. J. Taylor, Marcel Dekker, 1983.

6
Interfacing

Objectives

☐ To explain the requirements for interfacing circuitry in terms of electrical signal levels, and timing.

☐ To review the input–output arrangements used in microprocessor-based systems.

☐ To describe the operation of specialized interfacing chips.

☐ To consider the methods available for the transfer of data over a greater distance.

☐ To introduce common interface standards as used in data logger and telemetry systems.

There are several key considerations when interfacing subsystems. These are that the electrical circuitry on the two sides of the interface must be compatible, so that the signals are not distorted during transmission across the interface; the system interconnections should not result in undue noise sensitivity, and the timing constraints of the two sides should be satisfied. In this chapter we consider the operating requirements of widely used digital logic circuits, as well as the more specialized devices commonly found in microprocessor-based systems, and see how these have led to the development of standard interfacing arrangements.

Digital circuitry

At least some part of any modern control or measurement system makes use of digital circuitry, and most problems arise in interfacing the analogue and digital sections. Once in the digital domain interfacing problems are far fewer but they can arise when we need to connect subsystems using dissimilar circuitry or timing methods. In order to interface such subsystems successfully it is important to appreciate the characteristics and limitations of the logic circuits used.

The first universally accepted family of integrated logic circuits was transistor–transistor logic (TTL), and it has developed through several generations since its introduction in the early 1960s. For well over a decade, TTL in its various forms was the dominant type of logic circuitry until displaced by NMOS microprocessor-based circuitry and dramatically improved CMOS technology. Many of the interfacing parameters in all types of logic circuit are, therefore, often quoted in terms of TTL performance. The standard commercial TTL family has the prefix 74 and the most popular forms in current use are Low-power Schottky (LS), Schottky (S), Advanced

See *Digital Logic Techniques* by T. J. Stonham.

Low-power Schottky (ALS) and Advanced Schottky (AS; also known as FAST). The output stage of all these variants is based on the totem-pole circuit shown in Fig. 6.1. The two transistors, T_1 and T_2, operate in push–pull and are driven in antiphase by signals derived from the input logic circuitry. Resistor R is included to protect the circuity during switching, when both transistors are conducting for a brief period, and to reduce the current surges on the +5 V supply, which are propagated as common mode disturbances to all gates fed from that supply.

When transistor T_2 is ON, the output voltage, V_o, drops to less than 0.5 V, and current is drawn in at the output terminal. When transistor T_1 is ON, the output voltage is pulled up to a value greater than 2.7 V, and current is provided from the output terminal. The typical output voltage in the high state, V_{OH}, is well below the +5 V of the supply because of the voltage dropped across the limiting resistor, R, and transistor T_1. TTL is designed as a current sinking logic, and the output current capability in the low state, I_{OL}, is much greater than that in the high state, I_{OH}. Typical figures are listed in Table 6.1.

Early forms of TTL gates were standardized to give a maximum output current, I_{OL}, of 16 mA and an I_{OH} of $-400\ \mu A$. At the gate input the currents needed to give correct switching were -1.6 mA in the low state and $40\ \mu A$ in the high state. The ratio I_{OL}/I_{IL} and I_{OH}/I_{IH} is, therefore, 10, which is referred to as the **fanout** of the gate and indicates that the gate can successfully drive up to ten similar gates. In many cases the loading and drive capabilities are quoted in terms of unit loads (UL), normalized to these standard TTL figures. Thus the high-state unit load is $40\ \mu A$, and the low-state unit load is 1.6 mA. If, by way of example, we take the figures for 74LS TTL we see that $I_{IL} = 0.36$ mA giving a unit load factor in the low state

Fig. 6.1 TTL totem-pole output stage

Table 6.1

	$V_{OH}(V)$		$V_{OL}(V)$		$I_{OH}(\mu A)$	$I_{OL}(mA)$	$V_{IH}(V)$	$I_{IL}(V)$	$I_{IH}(\mu A)$	$I_{IL}(mA)$
	min	typ	typ	max	max	max	min	max	max	max
74LS	2.7	3.4	0.35	0.5	-400	8	2	0.8	20	-0.36
74S	2.7	3.4	0.35	0.5	-1000	20	2	0.8	50	-2
74ALS	2.7	3.4	0.35	0.5	-400	8	2	0.8	20	-0.4
74AS	2.7	3.4	0.35	0.5	-1000	20	2	0.8	20	-0.6

of 0.255 UL (0.36/1.6). This is normally rounded to 0.25 UL. In the high state the load factor is 0.5 UL, (20 μA/40 μA). The output current, I_{OL}, is 8 mA so the low-level drive factor is 5 UL, (8.0 mA/1.6 mA), and the high-level drive factor is 10 UL, (400 μA/40 μA).

Calculate the unit load factors and drive factors for 74S, 74ALS and 74AS TTL.

Exercise 6.1

The totem-pole output circuit used in TTL provides active pull-up as well as active pull-down of the output voltage, and this gives very fast switching between states. In certain circumstances, however, the active pull-up operation is impracticable and various combinations of open-collector gates are available in which the upper half of the totem-pole circuit is omitted. It is then necessary to provide the collector load resistance externally. One major use of the open-collector gate is in wired-OR distributed logic where the outputs of several gates are connected together. The common point is held at the low voltage if the output transistor of any of the connected gates is switched ON, giving a negative logic OR function, or, conversely, the common point voltage rises only if the output transistors of all the connected gates are OFF, giving a positive AND function.

This is also known as 'dot AND' or 'phantom OR'.

When open-collector gates are used, I_{OH} represents the transistor leakage current flowing in to the turned off collector

A single resistor is necessary to provide the collective load, and its value must be chosen to satisfy two requirements: when all the connected transistors are OFF the output voltage, V_{OH}, must be maintained, and when only one of the transistors is ON it must be able to satisfy the output current demand. Thus, when the transistors are OFF:

The output will rise in accordance with the equation $V_o = V_s (1 - e^{-t/CR})$ where R is the collector resistance and C is the capacitance associated with the common connection. CR is the time constant.

$$R_{max} = \frac{V_{cc(min)} - V_{OH}}{nI_{OH} + m \times 40\ \mu A}$$

$$R_{min} = \frac{V_{cc(max)} - V_{OL}}{I_{OL} - m \times 1.6\ mA}$$

where n is the number of open-collector outputs connected together (Fig. 6.2) and m is the number of unit loads being driven from the common point.

Open-collector gates are also used as bus drivers, and negative logic convention should then be adopted.

Three 74LS open-collector gates are connected to give a wired-OR function which is fed to a similar gate. What collector resistor value should be used?

Worked example 6.1

Fig. 6.2 The wired-OR arrangement

I_{OH} for the open-collector gates is $-100 \mu A$, and $V_{cc} = 5 \pm 0.25$ V.

In this case $n = 3$ and $m = 1$, so

$$R_{max} = \frac{4.75 - 2.7}{3 \times 100 \mu A + 1 \times 40 \mu A} = \frac{2.05}{0.34} \times 10^3$$

$$= 6.03 \text{ k}\Omega$$

$$R_{min} = \frac{5.25 - 0.5}{8 \text{ mA} - 1 \times 6 \text{ mA}} = \frac{4.75}{6.4} \times 10^3$$

$$= 0.74 \text{ k}\Omega$$

The value must, therefore, lie between 0.74 kΩ and 6.03 kΩ. A low intermediate value, such as 1 kΩ, would give good speed but high dissipation, whereas a high intermediate value, such as 5.6 kΩ, would reduce the dissipation but also the speed.

Three-state (or tri-state) versions of TTL circuits are designed for use in bus-organized systems, where any one of several gates feeding a common bus line may be selected, but all the others meanwhile must be disabled. In the selected, or enabled, condition, the gate output circuitry behaves as a normal totem-pole circuit, switching the output voltage to either V_{OH} or V_{OL} dependent on which transistor is held on by the logic conditions at the input. When disabled, however, internal circuitry ensures that both transistors in the totem pole are switched off, regardless of the input logic, and the output appears as a high resistance, effectively disconnecting the gate from the bus and allowing control to be taken by one of the other gates. If all gates connected to the bus are switched to the third state, that is disabled, the output voltage is determined by gate leakage currents and usually settles at about 1.5 V.

Many other forms of TTL drive circuit are available for specialized applications. The 74LS244 octal bus driver, for example, is particularly useful in microprocessor systems where buses are normally multiples of eight bits wide, and is designed to provide the high current drive capability necessary to cope with the capacitive loading of a backplane bus or ribbon cable. It has the following specification:

$$I_{IH} = 20 \mu A \text{ max} \qquad I_{OH} = -15 \text{ mA max}$$

$$I_{IL} = -0.2 \text{ mA max} \qquad I_{OL} = 24 \text{ mA max}$$

This indicates average levels of load factor, being 0.5 UL in the high state and 0.125 UL in the low state, but high drive factors of 375 UL in the high state and 15 UL in the low state. Where cables with characteristic impedances of about 50 Ω are to be driven, circuits such as the 74S140 quad 2-input positive-NOR line driver may be used.

Complementary symmetry MOS logic (CMOS) has been commercially available since 1968 but it is only in recent years that advances in manufacturing techniques have enabled it to challenge TTL in its speed of operation. Modern buffered CMOS in the 74HC series is comparable in speed with LS TTL but has a dissipation which is less than 20% of the TTL. CMOS is a voltage-controlled logic in which the gates are made up from complementary pairs of enhancement MOS transistors, one with an n-channel the other with a p-channel. The power supply requirements are far less demanding than for TTL, and V_{DD} can have any value between +3 and +18 V. The output voltage levels are within 0.2 V of V_{DD} and ground. Being voltage controlled, the fanout of a typical gate is limited only by the capacitance introduced by the driven gates and the interconnections, and for practical purposes is taken to be about fifty when driving other CMOS gates. In TTL terms the gate is capable of sinking 4 mA so has a drive factor of 2.5 UL. As with TTL, special bus driver circuits are provided, with 6 mA capability, and three-state versions are also available.

For TTL compatability the power supply V_{DD} must be +5 volts.

Many microprocessors and their support chips are fabricated in n-channel silicon gate depletion load MOS technology (NMOS) in which n-channel transistors are used both for switching and as dynamic loads. Again, however, they are designed to be compatible with TTL.

The Synertek SY6522 Versatile Interface Adaptor, VIA, has the following specification:

Exercise 6.2

$V_{OH} = 2.4$ V min	$I_{OH} = -100\ \mu$A min	$I_{IH} = 100\ \mu$A min
$V_{OL} = 0.4$ V max	$I_{OL} = 1.6$ mA min	$I_{IL} = -1.6$ mA max

This device is also supplied by Rockwell.

Show that the drive factors are 2.5 UL in the high state and 1 UL in the low state.

The fastest commercially available logic is still emitter-coupled logic (ECL), though its speed advantage over TTL and CMOS is now so small that its use is restricted to applications where speed is the overriding consideration. It is a non-saturating form of logic so its power dissipation is considerably higher than other logic types, and in order to make full use of the speed, ground planes must be used and all interconnections dealt with as transmission lines.

A related logic, Emitter Function Logic (EFL) is used in modern gate array design. See *Digital Bipolar Integrated Circuits* by M. I. Elmasry.

Specialized interfacing chips

In interfacing a microprocessor-based system with external circuitry it is necessary to match both electrical signal levels and their timing. Although standard logic levels, nominally +5 V and 0 V, are used in microprocessor systems, the speed of operation is much greater than most external equipment and the transient nature of the data and control signals leads to interfacing problems. The output data in the majority of systems, for example, is available on the bus for only a few microseconds at the most. To ease these problems, manufacturers have produced standardized input–output (IO) devices, some of the more sophisticated having timers, counters, on-board RAM and even ADC included. The most common standard IO devices are listed in Table 6.2.

These devices provide two or three parallel ports, compatible with TTL levels, and usually having edge-triggered capability when inputting. The characteristics are programmable, and each port can act as input or output,

Table 6.2

Manufacturer*	Microprocessor series	Device number
Intel	8085A	8255A (PPI)
Motorola	6800	6821 (PIA)
Rockwell	6500	6522 (VIA)
Zilog	Z80	8420 (PIO)

*A range of alternative sources is available.

or, in most cases, any mixture of inputs and outputs. The cost is only a few pounds and it is seldom worth constructing a special circuit if straightforward parallel data transfer is required. In transferring data from a digital transducer, such as an encoding disc, or from single inputs, such as limit switches or bi-metallic temperature sensors, a port can be programmed to act as an input port (Fig. 6.3).

Internal registers are used to control the operation of these devices and the registers must be set up correctly, using an initialization routine at the beginning of the program, before the ports can be used. This is not always a simple procedure. The 6522 Versatile Interface Adaptor (VIA), for example, has sixteen internal registers (including the data registers) which have to be set correctly. In practice, however, simple IO operations can be carried out using only a few of the registers, and in the VIA we need only the Data Register A (DRA) and a Data Direction Register A (DDRA). The second port, B, would involve DRB and DDRB. The data direction register controls the setting of the port bits as either input or output. Any bit of the DDR which is at '1' sets the corresponding bit of the port to act as an output, and, conversely, clearing any bit of the DDR to '0' switches the corresponding port bit to an input. When set to input mode, internal pull-up resistors hold the port bits high and each bit presents to the driving devices a loading equivalent to one TTL load. If programmed as an output, the port bits reflect the settings of the data register (DR), and each bit is capable of driving one TTL load. Whilst this is sufficient to drive a small relay, or possibly an LED,

The two ports, A and B, are not identical, the output circuits being as follows;

Port B can source up to 1 mA and is more versatile as an output port.

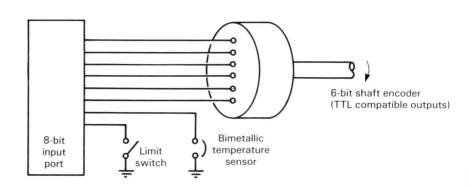

Fig. 6.3 Use of an input port

126

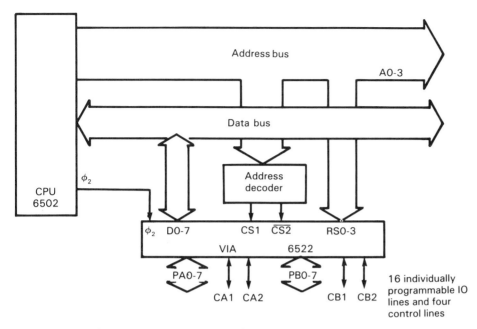

Fig. 6.4 Memory-mapped input-output arrangement

buffer amplifiers are often required, and, again, standard parts are available, both as TTL drivers and in Darlington driver form.

Figure 6.4 shows a typical memory-mapped input–output arrangement. The address decoder responds to a specific address present on the address bus, and in doing so selects the VIA by setting CS1 high and CS2 low. A small system with only one peripheral device might dispense with the decoder circuit and simply connect CS1 and CS2 to the address lines A15 and A14 respectively. The VIA would then respond to any address which had the most significant bits equal to '10'. In this case, remembering that the four least significant address bits are connected to the VIA internal registers and respond to the hexadecimal (hex) codes 0 to F, we say that the VIA is mapped to locations 8000 to 800F. We ignore the fact that the VIA would respond to many other addresses, since the decoding is not exhaustive, and make sure that such addresses never appear in a program. This partial decoding is very common as it does not need as many devices, but it cannot be used where a large amount of memory is required.

A small system is to use a VIA to interface the following devices:

Worked example 6.2

1. two LEDs;
2. one relay, operating at 24 V dc and with a coil resistance of 240 Ω;
3. one seven-segment display decoder/driver;
4. one limit switch and a temperature sensitive switch;
5. one 6-bit optical shaft encoder, TTL compatible.

This can be achieved, as shown in Fig. 6.5, by configuring port A of the VIA as an input port (all bits of DDRA set to '0'), and port B as an output port

127

Fig. 6.5 Use of a VIA in a small system

(all bits of DDRB set to '1'). If we assume that the microprocessor is a 6502, the initialization routine would include the following program segment:

    LDA #00

    STA DDRA   all 8 bits of DRA set as inputs

    LDA #00F

    STA DDRB   all 8 bits of DRB set as outputs

Port A is operating in input mode, so internal pull-up resistors effectively provide one TTL load and the switches can be connected directly. The shaft encoder is TTL compatible so can also be connected directly. Since the output port bits have limited drive capability, buffering is needed when driving the LEDs and the relay, and a Darlington driver package is used here. This includes the protective diodes used in limiting the back-emf across the relay. The seven-segment decoder/driver can be driven directly from the port, and the high-voltage, open-collector circuits of the 7446A allow the 24 V relay supply to be used to power the display as well.

See chapter 4.

As befits its name, the VIA has many additional features, including two 16-bit timer/counters and an 8-bit shift register. These can be used in interval timing, for generating pulse trains and in performing serial data transfers, but their setting up and control is complex. Being 16-bit, the counters cannot be loaded in a single operation; it is arranged, therefore, that the low-order 8 bits are set into an internal latch, and then, as the high-order bits are loaded into the counter, the low-order bits are automatically transferred from the latch. Counting down begins as the count is loaded, and, on reaching zero, an interrupt flag is set. The counting can be controlled either by the internal processor clock or by an external clocking signal applied to bit 6 of port B. One of the counters has two latches and, after counting down to zero, the data in the latches can be reloaded automatically so that counting continues and the device acts as a free-running pulse generator with controlled frequency and mark-to-space ratio. Data can be shifted out of the 8-bit shift register to an output pin under the control of one of the timers or of an externally provided signal, and many other outputs are possible by clever use of the control registers. However, the reader must refer to more specialized books and technical data sheets for further information.

See, for example, *Microprocessor Systems Engineering* by R. C. Camp, T. A. Smay and C. J. Triska.

Keyboard interfacing is a frequent requirement in microcomputer system design. If the application requires only a few keys, they can be dealt with as in the previous example by considering them as individual switches. For a hex keypad, however, or a large keyboard, often containing as many as 92 keys, more efficient techniques are available. Considering the hex pad array of Fig. 6.6, we can detect any depressed key if we connect the eight wires directly to a programmable port and use a simple interrogation program. The program must implement the following steps:

Set port bits A–D as outputs and bits W–Z as inputs.
Output the bits '0000' on A–D and read the input code on W–Z.
Reverse the port settings so that bits A–D are inputs and W–Z are outputs.
Output the previously acquired code on W–Z and read the input code on A–D.

We could output '0000' again, but on many microprocessors this requires more instructions.

Note that the configuring of port bits as inputs automatically connects pull-up resistances, so that reading an unconnected input will register a logical '1'. The port now holds an 8-bit code uniquely defining the depressed key. Assume, for example, that switch '6' had been pressed. The code received on W–Z is '1101', and the subsequent code on A–D is '1011', giving a composite code of '11011011'. A simple 'table lookup' exercise then leads to the actual code assigned to that key.

The problem of key bounce can be overcome by interrogating the keypad at least twice, with a delay of a few milliseconds between readings. Identical codes from successive interrogations indicate a genuine key press. However,

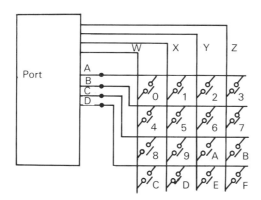

Fig. 6.6 Keypad encoding

Continual scanning is used both in calculators and in large multi-processor systems which use a dedicated microprocessor for that purpose.

one problem remains. Unless the microcomputer is to scan the keyboard continually some means is required to signal that a key has been pressed. This can be achieved by arranging that the port is configured in one particular mode while waiting for a keypress, say A–D as inputs. These inputs are also connected to a four-input AND gate, the output of which goes low when any key is pressed. This signal is then used to interrupt the processor.

It is often more economical to use specially developed keyboard-encoding chips. The MM74C922, for example, is a 4×4 keypad encoder with internal debounce circuitry, an output enable control, and a 'data available' flag which can be used to indicate to the processor that a key has been pressed. Such devices reduce the software demands on the processor, which is especially important in larger systems.

The interfacing of digital-to-analogue and analogue-to-digital converters to microcomputer systems is a very common requirement, and is made relatively simple by the widespread availability of converters specifically designed for such applications. If an analogue output is required, we could make use of the type of converter described in Chapter 4, such as the Ferranti ZN425E which gives an analogue voltage output directly from pin 14. The internal counter is disabled by setting pin 2, SELECT, to '0', and the input bits, which are TTL compatible, are connected directly to the microcomputer output port. A buffer amplifier can be used to adjust the analogue voltage to the appropriate level. However, in many applications a simpler converter such as the ZN429E is more suitable. This device consists of an $R - 2R$ ladder network and TTL/CMOS-compatible switches, and a typical arrangement is shown in Fig. 6.7. An external reference voltage is required, the maximum recommended value being $+3$ V, and generally a reference diode, such as the National Semiconductor LM136/336, is used, providing a defined voltage between 2 and 3 V. To calibrate the converter a buffer amplifier must be used. The 741 op-amp circuit shown is suitable, and also removes the offset voltage, typically 3 mV, which is present at the converter output when all the input bits are at zero.

The ZN429E must be coupled into the microcomputer system through a port, as it has no means of latching data itself, and in a very simple system the ZN428E (Fig. 6.8) may be preferable, since it can be connected directly

Fig. 6.7 Digital-to-analogue converter

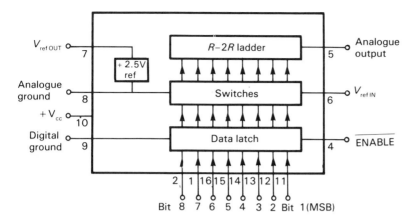

Fig. 6.8 The ZN428E digital-to-analogue converter

on to the system bus. The enable control signal, when taken low, allows data to be written into the latches and the data is held when the signal goes high. Thus the latches can take the place of the port and an 8-bit digital-to-analogue converter can be constructed as shown in Fig. 6.9. Now, valid data

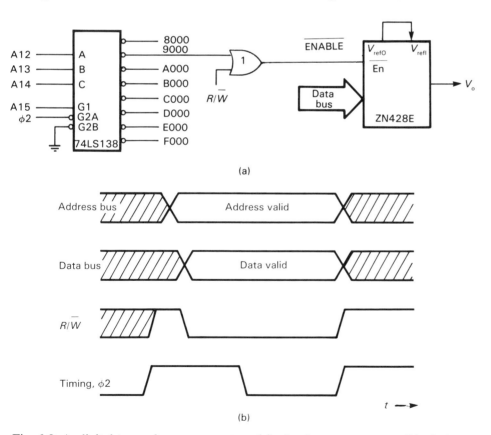

Fig. 6.9 A digital-to-analogue converter; (a) circuit arrangements (b) timing waveforms

131

is present on the microprocessor data bus only at certain specified times. With the circuit given, and assuming, by way of example, that the processor is executing the instruction STA 9000 (store the content of the accumulator in memory location 9000), the control and data waveforms would appear as in Fig. 6.9(b). The address-decoding circuitry generates an enable signal when the data from the accumulator appears on the data bus, allowing the data to be latched into the converter. To the programmer, the converter appears as a memory location which can be written into in the usual way. The ZN428E also has an on-board reference voltage which can be used to advantage, though a signal-conditioning amplifier is still required.

For the acquisition of analogue data an analogue-to-digital converter of some sort is needed. In many cases, where accuracy of the order of six to eight bits is sufficient and high speed is not essential, we can make the microcomputer do all the work with very few external components. For example, using the ZN428E converter and a comparator, as in Fig. 6.10(a), we arrive at a circuit giving seven- to eight-bit accuracy. Its operation is summarized in the flow diagram of Fig. 6.10(b), and is as follows. During initialization, bit 0 of port A is configured as an input and port B as an 8-bit output port. This port is set to zero. The unknown voltage is applied to the comparator input and, as the port value is incremented, the resulting voltage from the digital-to-analogue converter eventually exceeds the unknown voltage and the comparator changes state. At this point the value held in the port is the digital equivalent of the unknown voltage, to within the resolution of the circuit, which in this example is 1 part in 256.

This method is adequate in simple applications but it is usually more cost-effective to employ a specialized converter chip. The Analog Devices AD574, for example, is a 12-bit analogue-to-digital converter specifically designed for use in microprocessor-based systems. The 12-bit output is multiplexed on to a standard 8-bit data bus, and START and End of Conversion (EOC) signals are used to control the operation. Conversion is initiated by writing to a specified location, which generates the START pulse. The EOC signal is used to interrupt the processor, and the two 8-bit bytes prepared by the converter are then transferred to the processor.

Figure 6.11 is an example of a practical weighing system using the AD574. In this example, a strain gauge load cell (typically of the type shown in Fig. 2.11) is used in conjunction with the SGA100 (RS 308–815) strain gauge amplifier to provide the analogue signal for the ADC. The output from the AD574 is interfaced to a microcomputer system through a parallel port device, in this case a 6522 VIA. Considering Fig. 6.11(a), the load cell strain gauge bridge derives its power from the amplifier unit through two emitter follower transistors used to boost the power capability of the amplifier drive circuit. There is a major problem with the bridge circuit configuration generally used with resistive strain gauges. This is that the common mode voltage existing at the bridge outputs is half that of the bridge supply voltage and unless equal positive and negative supplies are used, this voltage can cause severe overloading problems. The sensitive dc amplifier needs to respond to the tenths of a millivolt signals which appear across this common mode voltage. Driving the bridge with equal and opposite supply voltages ensures that the common mode voltage is zero. The SGA100 device is a

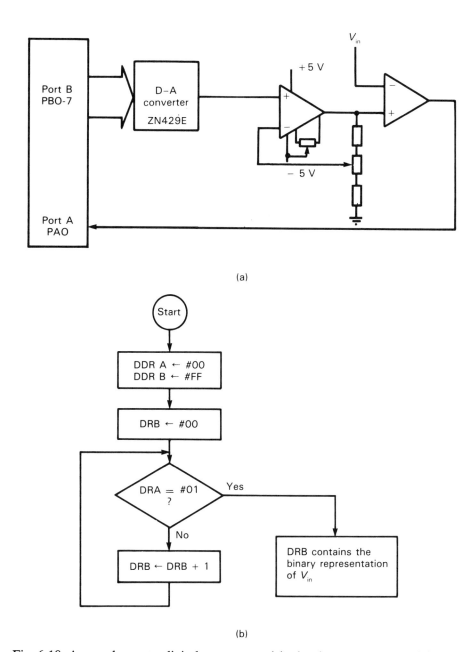

(a)

(b)

Fig. 6.10 An analogue-to-digital converter; (a) circuit arrangements (b) control flow diagram

special-purpose low-noise low-drift hybrid dc amplifier designed for use with resistive strain gauges. This particular amplifier unit removes the common mode voltage by controlling the negative bridge supply in such a way that the voltage at the negative signal input terminal is always zero. Thus for a symmetrical bridge the negative supply voltage is equal and opposite to the positive supply and the common mode voltage is zero.

(a)

* Calibration resistor RS 158–XXX

(b)

Fig. 6.11 Digital weighing system

The 'compensation' connection to the positive bridge supply point is used to monitor the bridge supply voltage. Very little current flows through this wire and thus the effect of voltage drops along the actual supply leads due to the bridge current can be ignored. Bridge supply volts (BS) and bridge offset (SZ) 'setzero' are adjustable. The 1N827 diodes (temperature-compensated zener diodes) are used to provide a stable voltage reference source for this purpose. The amplifier is set to a gain of 750 which, in this case, gives a signal range after threshold adjustment of 0 to 10 V. The output from the amplifier is taken to a sample/hold device, the LF398. This device is controlled by the computer, the same signal initiating a conversion cycle from the ADC, conversion taking place during the 'hold' time. The 'hold' capacitor has a polypropylene dielectric. The quality of this capacitor is crucial (see Chapter 4). Care must also be taken to ensure that the input to the LF398 is adequately shielded from noise signals, in particular from the sample/hold pin, and a 'guard ring', as discussed in Chapter 3, is recommended.

The AD574 is operated in its 12-bit parallel mode, the four most significant bits of the data are multiplexed with bits D4–D7 of the least significant eight bits of the data. Multiplexing is controlled by the least significant address line, A0. Selection of the converter is via pin 3, the selection signal being derived from an address decode unit. In operation, the conversion unit appears to be located at two consecutive addresses in the microcomputer address space. Conversion is started by writing to the converter and after a suitable delay (in this application the EOC signal is not used) reading of the two addresses yields the 12-bit digital equivalent of the analogue input signal.

Note that provision is made for multichannel operation in that the signals to the sample/hold circuit are fed from an LF13201 analogue switch which is controlled from the computer port. This will allow any one of four channels to be selected. Provision for checking the calibration of the system automatically is provided. This is done by placing a known-value resistance across one of the bridge arms. The change in signal is then compared with that calculated (and expected). The resistor used must be of very high stability. In this example a precision wirewound resistor is switched across the bridge arm using a reed relay unit. By taking readings of the bridge output before and after the resistor is switched in, the computer can calibrate the system, taking into account temperature drift for example, before each weight measurement.

Transfers of data over greater distances

The interfacing arrangements considered so far have dealt mainly with subsystems in close proximity. In many cases, however, it is necessary to transfer data between one system and a more remote system, and various techniques have been developed to deal efficiently with specific sets of circumstances. In longer distance interfacing the designer or user is normally in no position to modify the remote equipment, and it is very important that its specification and characteristics are well defined, so that the interfacing circuitry can match in terms of electrical signal requirements, timing and physical connections. Over recent years, therefore, several international standards have evolved and, because of the continuing rapid changes as

"The nice thing about standards is that you have so many to choose from. Furthermore, if you do not like any of them you can just wait for next year's model." A. S. Tanenbaum, *Computer Networks*

Named after Emile Baudot (1845–1903), a French telegraph engineer. The baud rate is actually defined as the reciprocal of the shortest time period used in the code, and if multi-bit coding is used the bit rate is greater than the baud rate.

A byte can have any number of bits but is commonly taken as having eight bits.

'Intelligent' is used to indicate the presence of some computing power.

See O'Reilly chapter 3.

See O'Reilly again, line coding, chapter 7.

computer-based digital electronics and communications engineering converge, they will no doubt continue to evolve.

Many considerations arise in deciding which are the most appropriate techniques to use. The cost of the interconnecting channel varies dramatically; distances up to a few hundred metres can use twisted pairs or coaxial cable. Medium and longer distances can use leased lines of the telephone network, or the Integrated Services Digital Network (ISDN) which is being introduced. Higher volumes of traffic may justify the use of optic fibres. Telemetry systems often use radio links and if necessary, of course, communications satellite channels are available. In deciding on a particular medium, the designer must balance the cost against the speed and accuracy of data transmission required. There are several fundamental choices which affect these decisions, the major ones being between serial and parallel operation, duplex and half-duplex channels, and synchronous and asynchronous timing.

Serial transmission uses a single path to carry successive information bits as a time-separated sequence. It is, in fact, a form of time-division multiplex (TDM). The eight bits of a standard code character, such as ASCII or EBCDIC for example, are sent along a single line in eight successive time intervals. In most digital transmission systems the period of each bit is constant, and the inverse of the bit time period is known as the *baud rate*. In parallel transmission, the eight bits of the character are sent simultaneously on eight separate paths, and it is complete characters that are time separated on the channel. This method is therefore sometimes referred to as bit parallel/byte serial. The parallel paths can be physically separate circuits, as is the case in the bus-organized systems using microprocessors and also in several standards designed for transfer of data between intelligent instruments. In many longer distance communication systems using radio and telephone circuits, however, the information on the parallel paths is frequency division multiplexed so that a single broadband channel carries all the information, with each path restricted to a specific narrower range of frequencies within that band.

In some cases it is sufficient to transmit information in one direction only: from a central station to outlying units, perhaps, or from remote sensing equipment to a central control unit. This 'A to B' transmission is known as simplex operation. In most cases, however, some two-way traffic is necessary, either to deal with the type of data involved or to allow acknowledgement of receipt of data, (and, by extension, a request for data to be repeated if the received data has been corrupted in transit). Half-duplex operation involves the use of a reversible channel which allows data flow 'from A to B' and 'from B to A' but not simultaneously. Simultaneous transmission in both directions, as for example when carrying speech on the telephone network, is known as full-duplex operation.

Since digital data is always time dependent, it is necessary to ensure that the circuitry at the two ends of a transmission channel operate in synchronism. Synchronous transmission makes use of a regular clocking signal which is used to keep the timing circuits at each end in step. This clocking waveform can take the form of a signal sent over a separate line, or it can be combined with the data by use of self-clocking codes. In many cases, special synchronizing characters (SYNC) are transmitted at regular intervals of about a second.

The serial data arriving at the receiver is shifted into a buffer register and the most recently received eight bits are compared with the SYNC code. Recognition of the SYNC character enables the receiver timing circuits to 'pull' into step and maintain synchronization. In order to protect against the possibility of a SYNC character occurring at random in the data stream, many systems require two consecutive SYNC characters, and this is known as bisync control.

Asynchronous operation is used in slower transmission systems and relies on timing information carried with each character. The data bits which make up the character to be transmitted are framed by START and STOP bits, as in Fig. 6.12. When the line is active but not carrying data, it sits at the upper, or idling, level. The start of every group of transmitted bits is a negative-going edge, from which the timing of all the subsequent data bits in the group is derived.

Special chips are available to carry out the conversion between the parallel form of data required by a microprocessor-based system and the serial form carried by the transmission channel, and to deal with the framing of each character. This type of chip is usually referred to as a Universal Asynchronous Receiver/Transmitter (UART), though Asynchronous Communications Interface Adaptor (ACIA) and several other descriptions are also used. The basic structure of a UART is shown in Fig. 6.13, the four main sections being the receiver, transmitter, control and modem control. Signals to and from the microprocessor are shown on the left, and those to and from external circuits are shown on the right. Both receive and transmit sections make use of double buffering to allow faster speed of operation – in some cases up to 500 kilobaud, although 19.2 kilobaud is a more common operating maximum. Consider first the transmitter section: the T×RDY signal is used to indicate that the UART is ready to accept a character for transmission, even though a previous character may still be shifting out of the shift register. The new character provided by the microprocessor is transferred from the data bus into the transmit buffer and, as soon as the previous character is shifted out, it is loaded into the shift register for serialization. The transmitter control now indicates on T×RDY that it can accept another character. If both the buffer and the shift register are empty, the T×E signal is set to indicate that the transmitter is empty. The shift register is controlled by the externally generated clock signal T×C, and the data bits of the character are shifted out least significant bit first, automatically preceded by a start bit and followed by a parity bit (if required) and the stop bit or bits. The UART must be programmed to deal with the correct number of data bits in each character, to include the parity bit (either odd or even) when required, and to add one or

Modem is a contraction of modulator-demodulator.

Start bit | Eight data bits | Stop Bit(s)

Fig. 6.12 Asynchronous character framing

Key: Tx Transmitter
    Rx Receiver
    S_T Stop bit
    S_P Start bit
    P Parity check bit

Processor signals: CLK   Clock
    RST   Reset
    CS   Chip select
    C/D   Control/data word
    RD   Read
    WR   Write
    TxE   Transmit buffer empty
    TxRDY Transmitter ready for data
    RxRDY Receiver ready with data

Modem signals:
    DSR Data set ready
    CTS Clear to send
    DTR Data terminal ready
    RTS Ready to send

UART signals:
    TxD Transmit data
    TxC Transmit clock
    RxD Receive data
    RxC Receive Clock

Fig. 6.13 The universal asynchronous receiver/transmitter

two stop bits. This information is provided by the user immediately after initializing the system, in the form of a control byte written into the UART control register.

The receiver section acts as the serial-to-parallel converter for incoming data. The first indication to the receiver that a character is on the line is that

138

the line voltage goes low as the start bit arrives. Unfortunately, the signals present on the line, in practice, are far removed from the idealized waveforms we use to represent them, and many of the spikes and other transient noise signals we pick up could appear as the beginning of a START bit. The UART must therefore contain circuitry to detect and reject false starts. This is achieved by double checking each START bit, using a clock frequency several times greater than the data baud rate. Assume the clock rate gives 16 pulses per data bit. When a possible START bit is detected, 8 clock pulses are counted and the input is then sampled again. If the level is still low, a valid START bit is assumed and the following data bits are shifted in on successive multiples of 16 pulses. By this means the data bits should be sampled at the centre of each bit period, and a wide difference in bit timing between transmitter and receiver can be tolerated without loss of data (Fig. 6.14). Each new character redefines the timing, and the STOP bit following each character gives time for the receiver to recover if its clock is running slower than the transmitter's clock. When the incoming data bits have all been shifted into the register, the byte is transferred to the receive buffer and the R×RDY signal is generated to indicate to the processor that a character has been received. This character must be read into the processor before the next incoming character is complete, otherwise the first character is overwritten by the second. In such an event, an overrun flag bit is set in the status word to indicate that a byte has been lost. Other flag bits are set if a parity failure is detected or if a framing error, such as a missing STOP bit, is encountered.

The transmit buffer and control register are both write only; the receive buffer and status register are both read only. In order to select a particular register, therefore, the processor must indicate either control or data, using the C/D̄ line, and make use of the correct read or write signal as summarized in Table 6.3.

The UART inputs and outputs are all TTL compatible and, over very short distances, standard TTL drivers and receivers can be used. In noisy electrical environments, however, it is necessary to use special circuits and techniques. Noise signals induced from adjacent circuits, or even slight variations in the earth reference, due to heavy currents in the common earth wire, can cause problems in recovering the transmitted data. Use of twisted pairs of wires, or of coaxial cable with the outer sheath earthed, may allow unbalanced drivers

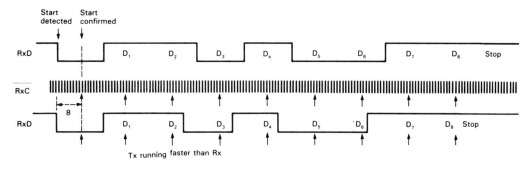

Fig. 6.14 Generation of timing from the START bit

Table 6.3

$\overline{CS}$	$C/\overline{D}$	$\overline{RD}$	$\overline{WR}$	Register selected
1	X	X	X	UART not selected
0	0	1	0	Transmit buffer
0	0	0	1	Receive buffer
0	1	1	0	Control register
0	1	0	1	Status register

bps stands for bits per second.

and single-ended receiver circuits (Fig. 6.15a) to operate satisfactorily at up to 20 kbps over a distance of about 20 m. Longer distances and higher signalling rates require balanced operation (Fig. 6.15b) in which the line is driven differentially. If noise voltages are induced they appear in common mode, so they are rejected at the receiver. The earth connections are not now part of the signal circuit, so earth currents will not affect the performance. Distances in excess of 1 km are possible with balanced operation, and if the cable is terminated in its characteristic impedance (usually between 50 and 150 Ω), so as to prevent reflections, speeds of 10 MHz can be achieved.

At high speed, or over long distances, where the edge speeds of the digital signals are less than the electric length of the cable, it must be considered as a transmission line. Remember that an electrical signal travels at (approximately) $\sqrt{\varepsilon_r}$ ns per foot! ε_r is the relative permittivity, typically between 2 and 4 for common insulators and printed circuit boards.

In driving solenoids, such as those commonly used in teletypewriters, a current drive provides faster operation. In this case, a current loop is required. Several current levels have achieved popularity, but the most popular is the 20 mA current loop, apparently because, in earlier days, 20 mA was found to be a good level for keeping switch contacts clean. The currents can be ±20 mA, in polar operation, or, more normally, +20 mA (indicating logic '1') and zero current (indicating logic '0'), in neutral working. Most modern applications in which current loops are used to give isolation, or to

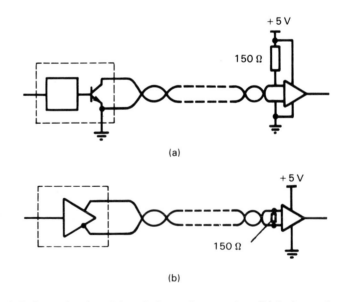

(a)

(b)

Fig. 6.15 Driver circuits; (a) unbalanced operation (b) balanced operation

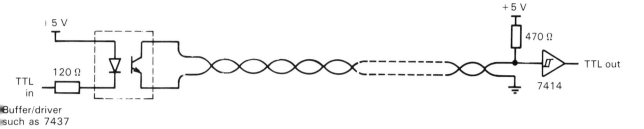

Fig. 6.16 Use of an opto-isolator to achieve electrical isolation via a ten milliamp current loop

provide noise immunity, make extensive use of opto-isolators, as illustrated in Fig. 6.16. In these systems the current need not be 20 mA, and can be chosen to suit the sensitivity of the opto-isolators.

Interfacing standards

One of the standards most widely accepted internationally is the American Electronic Industries Association (EIA), RS232-C standard for serial data. The 'C' indicates the most recent revision, published in August 1969. An almost identical standard is issued by the European CCITT (International Telegraph and Telephone Consultative Committee) with the number V24.

The standard was devised to deal with serial transfer of data over considerable distances, making use of existing networks such as the public telephone system. It is, therefore, defined in terms of data transfer between Data Communication Equipment (DCE), and Data Terminating Equipment (DTE), as illustrated in Fig. 6.17. The DCE is commonly a dataset, or modem, which converts the ON/OFF switching of the digital circuits to the frequency shift keying (FSK) methods used on the telephone channel. The DTE is often a terminal. The same standard can be, and is now widely used in interconnecting digital equipment quite independent of the telephone network.

See O'Reilly chapter 7 again.

Mechanically the standard specifies a 25-way D-type connector and defines the signal to be carried by each of the pins. The connections are divided into four categories:

So called because of its 'D' shape.

Data signals, both transmitted and received;

Fig. 6.17 Serial transfer of data over the telephone network

Control signals to provide orderly transfer of the data;
Timing signals;
Ground connections.

The complete list is given in Appendix A, but the vast majority of applications use only a very small subset of the signals. These are the data signals,

T×D Transmitted data (pin 2)
R×D Received data (pin 3)

and the control signals

CTS Clear to Send (pin 5)
RTS Ready to Send (pin 4)
DTR Data Terminal Ready (pin 20)
DSR Data Set Ready (pin 6)

A common signal earth is provided using pin 7.

The control signals are used in a 'handshake' convention, which requires that any control operation initiated by one side in setting up a transfer must be acknowledged by a control signal from the other side before the transfer can proceed further. Thus, suppose a terminal, the DTE, wishes to transmit data via the Dataset, the DCE, then the following handshaking sequence must be completed before the DTE is allowed to transmit:

The DTE sets DTR (pin 20) to ON and awaits a response from the DCE.
The DCE, if ready, sets DSR (pin 6) to ON.
The DTE now sets RTS (pin 4) to ON and awaits a response from the DCE.
The DCE, when ready for data, sets CTS (pin 5) to ON.
The DTE transmits its data on T×D (pin 2).

We see that the DTE is permitted to transmit data only when the ON condition is present on all the control lines DTR, DSR, RTS and CTS.

Electrically the standard specifies ranges of open-circuit voltages which are acceptable, shown in Fig. 6.18, but when the lines are terminated the voltages

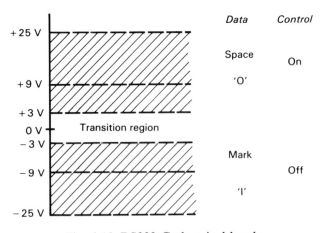

Fig. 6.18 RS232-C electrical levels

lie between 5 V and 15 V. Many standard driver and receiver circuits, such as the Motorola MC1488 and MC1489, interface directly with TTL circuitry. The MC1488 requires ±9 V supplies to provide the RS232-C levels, but otherwise behaves as a quad TTL gate package. The need for a negative supply is a considerable embarrassment in modern equipment which relies for all other purposes on a single positive supply voltage. With this in mind, and taking advantage of higher signalling rates over greater distances, RS232-C is gradually being replaced by a new standard, RS449. This must be read in conjunction with standards RS422 and RS423 which define the electrical characteristics of balanced voltage and unbalanced voltage digital interface circuits respectively. Signal voltages, either between balanced lines or between the signal line and earth, are required to lie between +4 V and +6 V for the SPACE, or logical '0', and less than 200 mV for the MARK, or logical '1' condition.

Also available as SN75188 and SN75189.

In many industrial control and telemetry systems the standards which evolved primarily for data transmission do not provide sufficient performance or flexibility, and a more recent introduction by the Intel Corporation is the Bitbus interconnect. This is a serial bus optimized for high-speed transfer of short control messages between as many as 250 nodes, if required, and at data rates up to 2.4 Mbps. Simple twisted-pair cables are used as the transmission medium to minimize cost, and the use of 'open' standards for both the physical level and the message-passing software ensures compatibility with systems from any supplier.

National Semiconductor has a somewhat similar arrangement known as the Microwire Plus interface.

Several parallel interface standards have been established for some time. The IEEE-696 interface is widely used in modular systems, but the major interconnection standard between digital instruments is the IEEE-488, or General Purpose Interface Bus (GPIB). As with RS232-C, a particular connector is specified, but the protocols governing the setting up of a transaction are much more complicated. The process is a handshaking procedure between the transmitter, the *talker*, and a receiver or receivers, the *listeners*, which allows 8-bit data bytes to be transferred asynchronously at the fastest rate possible dependent on the equipment involved. The standard specifies 24 lines on the bus and these are made up of 8 data lines (DIO1–8), 5 bus management lines and 3 data transfer control lines, with 6 ground return lines allocated to specific command lines, a general ground line and a ground shield connection making up the set.

Originally known as the S-100 bus.

A useful reference here is the Motorola MC68488 GPIA User's Manual, MC68488UM(AD), 1980.

Not DIO 0–7 as usual.

There are essentially two levels of operation on the bus: the *physical* level covers the operation of the bus signals themselves in actually transferring messages along the cables between units. The *protocol* level consists of the set of rules governing how the transactions are set up and managed. In general, the controller, or *bus-master*, assumes complete control and is responsible for setting up the talker and listeners for each transaction. Suppose, for instance, we have a system in which the IEEE bus connects several instruments, including perhaps a digital voltmeter, a printer, a programmable power supply unit and the computer, which acts as the system controller. When we switch on, the controller initializes the system by sending an *interface clear* (IFC) signal to all interface units, followed by a coded message, *device clear* (DCL), to set all units to a known quiescent state. The controller then has to set up, or program, the units by sending a sequence of messages to each in turn. For each unit we need a *listen address* message to alert the unit, then the necessary programming data word and finally an *unlisten address* to tell it to stand down again. At some stage we may wish to send data from the digital voltmeter to the printer, so the transaction must be set up by the controller. The first thing is for the controller to send the listen address to the voltmeter, followed by the data word needed to activate its digital output circuitry, and then the unlisten command. Next it sends the listen address message for the printer, and follows that with the talk address message to the voltmeter. The 'route' is now set up and the voltmeter feeds a succession of data words on to the bus and the printer accepts them from the bus. An **end of data string** signal is fed on to the bus when the last byte of the data is transmitted, to indicate that the transaction is complete. The protocol sequences are specified in terms of bus signals, such as the interface clear (IFC) signal, and bus messages, such as the listen address, talk address, data word, unlisten command, and so on. The messages are carried on the data lines with valid data being indicated by an active control line *data valid* (DAV).

The signals on the data lines are interpreted as an 8-bit byte of data unless the attention (ATN) line is at logical '1', in which case they are intended as a 7-bit command. The command is usually from one of the four classes indicated by bits 6 and 7 of the byte:

00 defines a universal bus management command

01 is a listener address

10 is a talker address

and

11 is a secondary address or command.

When an address is indicated, the actual value is held in bits 1 to 5, but the address with all ones is reserved as a universal unlisten or untalk command. Figure 6.19(b) shows the sequence to carry through the transfer of binary-coded data from the voltmeter to the printer.

The remaining bus management lines shown in Fig. 6.19(a), REN, SRQ and EOI, are used in rather more specialized operations. **Remote enable** (REN) can disable any local controls on an instrument, so putting it under bus control. **Service request** (SRQ) is an interrupt type signal, and **end or identify** (EOI), as well as indicating the last byte in a multiple-byte transfer is used by the controller when identifying interrupting units.

The electrical signals on the bus are specified in terms of TTL levels, using negative logic. Thus a signal between $+2$ V and $+5$ V is the high or '0' state, and a signal between 0 V and $+0.8$ V is the low or '1' state. The active low convention is used to allow wired-OR operation, and open-collector drivers are used in most cases, though three-state drivers can be used on the data lines and some of the control lines. The total bus length is restricted to 20 m, and each signal line must be correctly terminated by a resistive load, with a diode clamp to prevent negative voltage excursions. By using a terminated bus to reduce reflections, the data transfer rate can be taken up to 2 Mbytes per second, but the actual overall transfer rate is governed by the speed at which the slowest unit involved in the transfer can provide or absorb the data.

(a) Pin connections

Signal	Pin		Pin	Signal
DIO1	1		24	DIO5
DIO2	2		14	DIO6
DIO3	3		15	DIO7
DIO4	4		16	DIO8
EO1	5		17	REN
DAV	6		18	DAV
NRFD	7		19	NRFD
NDAC	8		20	NDAC
IFC	9		21	IFC
SRQ	10		22	SRQ
ATN	11		23	ATN
SHIELD	12		24	LOGIC GND

GND RETURNS (pins 18–23)

DIO1–8 DATA BUS
EOI END OR IDENTIFY
REN REMOTE ENABLE
DAV DATA VALID
NRFD NOT READY FOR DATA
NDAC NOT DATA ACCEPTED
IFC INTERFACE CLEAR
SRQ SERVICE REQUEST
ATN ATTENTION

(a)

(b) IEEE bus transactions for data transfer

	ATN	EOI	IFC	8	7	6	5	4	3	2	1	
IFC	0	0	1	0	0	0	0	0	0	0	0	Reset
DCL	1	0	0	0	0	0	1	0	1	0	0	Device clear
LA (Voltmeter)	1	0	0	0	0	1	0	0	1	0	0	Listen address
DAB	0	0	0	1	1	0	0	0	0	0	0	Voltmeter set-up code
UNL	1	0	0	0	0	1	1	1	1	1	1	Unlisten
LA (Printer)	1	0	0	0	0	1	0	0	1	1	0	Listen address
TA (Voltmeter)	1	0	0	0	1	0	0	0	1	0	1	Talk address
DAB	0	0	0	0	0	1	1	1	0	0	0	Data byte
DAB	0	0	0	0	0	1	1	0	1	0	1	Data byte
DAB	0	0	0	0	0	1	1	0	1	1	1	Data byte
⋮												
DAB/EOI	0	1	0	0	0	1	1	0	0	1	0	Last data byte
UNT	1	0	0	0	1	0	1	1	1	1	1	Untalk
UNL	1	0	0	0	0	1	1	1	1	1	1	Unlisten

(The header columns 8 7 6 5 4 3 2 1 are grouped under "DATA BITS".)

(b)

Fig. 6.19 (a) Standard pin connections for IEEE-488 bus signals (b) IEEE bus transactions for data transfer

The handshaking is controlled by the three signals **data valid** (DAV), **not ready for data** (NRFD), and **not data accepted** (NDAC), and proceeds as follows. The talker having output the current data bits and sensing that the listener is ready for data, because NRFD is high and NDAC is low, sets DAV low. The listener detects the change on the DAV line and so pulls NRFD low. When the data bits have been absorbed into the buffer, NDAC is allowed to go high. The talker maintains the data on the bus until it senses that NDAC has gone high, then sets DAV high and replaces the data with the next word. The cycle then repeats. The handshake removes the problems arising from the delays which occur when long interconnections are used; data will be accepted only when the negative edge of the signal DAV appears at the receiver and will be changed by the transmitter only when it, in turn, receives the negative-going edge of NRFD. The complete procedure is shown in Fig. 6.20. The active low wired-OR arrangement ensures that several listeners can be active at once, since the control line will remain low until the last listener is ready to let it go high. One unfortunate feature of any

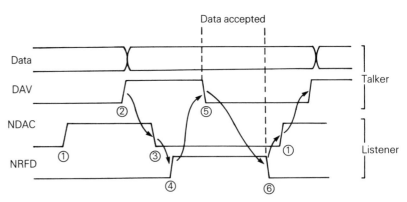

Fig. 6.20 Flowchart and timing waveforms of the IEEE-488 standard communication procedure, showing the 'handshake' principle

handshake method is that a transfer can 'hang up' if a response is not forthcoming and a *watchdog timer* must be built in to prevent an endless wait. As each signal is generated the timer is restarted and, if it 'times out' before a response is received, then the whole transfer is aborted.

The IEEE 696, or S-100, bus was developed as an interconnecting bus for 8080 and Z80 microprocessor-based systems but has been extended to cater for 16-bit processors as well. The name S-100 arises from the 100-line bus specified, though only 93 of the lines are actually defined and the remaining 7 may be allocated as required in a given application. The lines defined are:

16 data lines

16 address lines

8 extended address lines

8 priority interrupt lines

8 status lines – read, write, fetch, etc.

11 control signals

8 direct memory access (DMA) control signals

9 additional control lines – reset, clock, inhibit, etc.

9 power and ground return lines

Although well established, and still popular in certain corners of the market, the IEEE 696 bus has largely been displaced by other standard buses designed specifically to take advantage of the facilities available with 16- and 32-bit microprocessors. Unfortunately, individual manufacturers, aware of the desirability of a standard bus, nevertheless insisted on introducing standards which suited their own products with little thought for the customers' need to deal with alternative suppliers as well. We therefore have a range of standard bus offerings, but the two main contenders are the Multibus (versions I and II), introduced by Intel, and the VMEbus developed from Motorola's Versabus. Both these buses are well-structured and widely used throughout industry, and in recent years an attempt has been made to define a single bus which incorporates the best features of the two. This is the Futurebus, defined in IEEE standard 896. Futurebus is a 32-bit interface standard using a 96-way bus connector and allowing for the interconnection of up to 24 subsystem boards or modules. The connector specified is a 96-pin 3-row Euroconnector and as many as 22 of those pins are assigned to power supply connections (+5 V and ground) to help with the suppression of noise problems, which commonly arise when interconnecting subsystems.

A further five pins are used as high frequency noise shunts.

All bus transfers are carried out asynchronously with the handshaking again ensuring that transactions are completed as rapidly as possible. The speed of operation is limited by physical restrictions on the rate at which signals can propagate along the backplane, but stripline techniques and special drivers mean that transfer rates of over 100 Mbytes per second should be possible. In order to save on board space, and the number of transceivers and connector pins required, the address and data buses are multiplexed on 32 lines. Each line is 'pulled up' to the high level by a backplane terminating resistor and provides a wired-OR function, so that any one, or more, of the open-collector type drivers can assert, or drive the line, to the low level. The bus transceivers are specially designed for use with striplines and do not use normal TTL voltage levels. A signal switching at high speed, with rise and fall times of a few nanoseconds, generates frequency components up into the

Striplines use a track sandwiched between two ground planes. 40 Mbytes per second are readily obtained in most cases.

Transmitter/receivers used in bidirectional lines can be set to transmit or to receive as required.

See *Digital Signal Transmissions* B L Hart

gigahertz range, and the wavelengths of these components can be much shorter than the interconnection distances. The bus lines must, therefore, be considered as a set of transmission lines and general transmission line theory applies. It is becoming increasingly important to take these factors into consideration as clock rates in high-performance microprocessor systems continue to increase.

Summary

Interfacing is probably the cause of the most headaches for the system designer. Some of the most common problems likely to be encountered have been discussed, and methods of overcoming them outlined. Electrical noise (like the poor!) will always be with us, and a thorough understanding of the mechanisms whereby unwanted signals can gain access to our circuits and systems is essential if reliable, trouble-free systems are to be constructed. Equally, in interfacing to digital circuits, an understanding of the capabilities and limitations of the different logic families is important. Many specialized devices are available to deal with the control of the inputting and outputting of data from digital systems, in particular those that are microprocessor-based. When interconnecting systems which are widely separated, there are several standard techniques available, and it is always worth while adhering to the standards wherever possible.

Review questions

1. If the capacitance associated with the wired-OR connection in the example on p. 123 is 150 pF, what value of collector resistance should be used to give a time constant of 0.58 μs?
2. What is key bounce and how can it be overcome?
3. Define baud rate.
4. What is meant by the terms 'half-duplex' and 'full-duplex'?

Further reading

1. *Interfacing the BBC Microcomputer,* B. R. Bannister and M. D. Whitehead, Macmillan, 1985.
2. *Microprocessor Systems Engineering,* R. C. Camp, T. A. Smay and C. J. Triska, Matrix Publishers, 1979.
3. *Interfacing to Microprocessors,* J. C. Cluley, Macmillan, 1983.
4. *Computers and Microprocessors: Components and Systems,* A. C. Downton, Van Nostrand, 1984.
5. *Electrical Instrumentation and Measurement Systems,* 2nd edn, B. A. Gregory, Macmillan, 1981.
6. *How to Control Electrical Noise,* M. Mardiguian, Don White Consultants Inc., 1983.
7. *Telecommunications Principles,* J. J. O'Reilly, Van Nostrand, 1984.

8. *Modern Active Filter Design,* R. Schaumann, M. A. Soderstrand and K. R. Laker (eds). IEEE Press, 1981.
9. *Computer Networks,* A. S. Tanenbaum, Prentice-Hall, 1981.
10. *Digital Signal Transmission,* B. L. Hart, Van Nostrand, 1988.
11. *Data Communications and Computer Networks,* F. Halsall, Addison-Wesley, 1985.

Appendix A RS232-C Standard Signals and Pin Numbers

Interchange circuit	Pin number	Description	Gnd	Data		Control		Timing	
				From DCE	To DCE	From DCE	To DCE	From DCE	To DCE
AA	1	Protective ground	×						
AB	7	Signal ground/common return	×						
BA	2	Transmitted data, T × D			×				
BB	3	Received data, R × D		×					
CA	4	Request to send, RTS					×		
CB	5	Clear to send, CTS				×			
CC	6	Data set ready, DSR				×			
CD	20	Data terminal ready, DTR					×		
CE	22	Ring indicator				×			
CF	8	Received line signal detector				×			
CG	21	Signal quality detector				×			
CH	23	Data signal rate selector (DTE)					×		
CI		Data signal rate selector (DCE)				×			
DA	24	Transmitter signal element timing (DTE)							×
DB	15	Transmitter signal element timing (DCE)						×	
DD	17	Receiver signal element timing (DCE)						×	
SBA	14	Secondary transmitted data			×				
SBB	16	Secondary received data		×					
SCA	19	Secondary request to send					×		
SCB	13	Secondary clear to send				×			
SCF	12	Secondary rec'd line signal detector				×			

Solutions to Problems

Chapter 1

1. $216\,\Omega$
2. $4.5\,\Omega$, $92\,\Omega$
3. 11.1%, 21%, $9\,mV/Pa$
4. 2%, -20%
5. $0.2\%\,{-}20\%$

Chapter 2

2. $0.76:1$
3. 3
4. $2\,\Omega$
5. $280\,mV\,mm^{-1}\,V^{-1}$, $0.82\,mm$

Chapter 3

1. 10
2. $5.3\,k\Omega$
3. 10, $1.005\,V$, 0.5%
4. $-8\,V$
5. $12.2\,k\Omega$
6. $0.19\,ms$

Chapter 4

1. $-2.56\,V$
2. $2.25\,V$
3. $2\,\mu s$ (Note that resolution is defined as a percentage of full scale)
4. $6.5\,\mu s$
5. (a) 0000,1000,1100,1010,1011
   (b) 0000,1000,1100,1010,1001
   (c) 0000,1000,0100,0110,0111

Chapter 5

4. 20 Hz (Use anti-alias LPF before the sample-and-hold circuit, with cutoff at ~5 Hz.)
5. $SNR = 20\lg 314 = 49.9\,dB$ (Compare approximate value given by $(6n + 2)$, i.e. 50 dB.)

Chapter 6

1. $3.9\,k\Omega$

Index

Notes

Notes

Notes

Notes

Notes

Notes

Notes

Notes

Notes

Notes